Medical Biochemistry

Publication of this book was assisted
by a grant from the McKnight Foundation
to the University of Minnesota Press's program
in the health sciences.

Medical Biochemistry

Principles and Experiments

John F. Van Pilsum and Robert J. Roon, Editors

University of Minnesota Press Minneapolis

Published by the University of Minnesota Press,
2037 University Avenue Southeast, Minneapolis MN 55414.
Published simultaneously in Canada by
Fitzhenry & Whiteside Limited, Markham.

Printed in the United States of America

ISBN 0-8166-1344-3

Contents

Preface

The laboratory experiments in this manual are designed to introduce students to biochemical methods used in the clinical laboratory and to assist them in the comprehension of biochemical principles. The manual focuses on those biochemical principles that students can best understand by performing experiments. Most of the experiments involve procedures that are now routinely used in clinical chemistry laboratories or will be in the foreseeable future. The clinical significance and limitations of laboratory procedures are stressed.

The material in this biochemistry manual has been used extensively to teach first-year medical students at the University of Minnesota. In recent years a method of cooperative learning has been employed with great success in this biochemistry course. The results of our experiments with this type of teaching have been published. (Roon, R. J., Van Pilsum, J. F., Harris, I., Rosenberg, P., Johnson, R., Liaw, C., and Rosenthal, L. 1983. The Experimental Use of Cooperative Learning Groups in a Biochemistry Laboratory Course for First-Year Medical Students. *Biochemical Education* 2:12.)

The following authors are members of the faculty of the Department of Biochemistry, the Medical School, the University of Minnesota, Minneapolis: John F. Van Pilsum, Robert J. Roon, Marilyn H. Koenst, John D. Lipscomb, James B. Howard, Ivan D. Frantz, Jr., Denise M. McGuire, Howard C. Towle, Dennis M. Livingston, Ronald D. Edstrom, Frank Ungar, Maureen A. Scaglia, James F. Koerner, and Charles W. Carr. Esther F. Freier is on the faculty of the Department of Laboratory Medicine and Pathology, and Robert P. Chandler is in the Department of Nuclear Pharmacy—both in the Medical School of the University of Minnesota.

Acknowledgments

The editors thank the following for reviewing the manual and making suggestions that have been incorporated into this publication: Thomas M. Devlin, Professor and Chairman, Department of Biological Chemistry, Hahnemann Medical College and Hospital, Philadelphia, Pennsylvania; Murray Saffron, Professor, Department of Biochemistry, Medical College of Ohio at Toledo, Ohio; and Arthur A. Spector, Professor, Department of Biochemistry, College of Medicine, State University of Iowa, Iowa City, Iowa.

The editors are grateful to Diana Randall, Research Fellow, Office of Curriculum Affairs of the Medical School, University of Minnesota, for assistance in the revision of the manual. We also thank the following University of Minnesota faculty members for their help with the implementation of the cooperative learning method in our biochemistry laboratory for first-year medical students: Pearl Rosenberg, Assistant Dean of the Medical School; Ilene Harris, Research Associate, Office of Curriculum Affairs of the Medical School; and Roger Johnson, Professor of Curriculum and Instruction.

Safety Precautions in the Laboratory

Every effort has been made to minimize the hazards in this laboratory. Nevertheless, some dangers remain, and the student should be aware of them and should understand the precautions that must be taken.

Obvious Safety Practices

- Do not perform any unauthorized experiments.
- Do not eat, drink or smoke in the laboratory.
- Know the location and operation of all safety equipment.
- Report any incidents immediately to the teaching assistant.

Safety Equipment

The student should know the location and proper operation of the following:

- shower
- eyewash
- fire extinguisher
- fire blanket
- fume hood

Mouth Pipetting

This is perhaps the most important aspect of safety in this laboratory, where blood plasma and other biological samples are often used. These samples must not be pipetted by mouth under any circumstances to avoid possible transmission of hepatitis. In addition, the student should wash his or her hands after working with biological samples. Most of the other solutions in the laboratory can be safely pipetted by mouth if pipetting is done with care. However, it is best to get into the habit of not pipetting by mouth.

Radioisotopes

Some of the experiments involve the use of radioactive substances. The level of radioactivity is very low, and the substances are quite harmless if handled with care. Like blood samples, radioactive samples must not be pipetted by mouth, and hands should be washed after using these samples. If radioactive material is spilled, the teaching assistant should be notified immediately.

Fire

If the fire alarm sounds, all students should leave the building immediately.

Broken Glassware

Broken glass should be cleaned up immediately. The pieces of broken glass should be placed in a container specified for that use and the area cleaned thoroughly to prevent injury. The student should seek attention immediately for any cuts.

Disposal of Solutions

In most cases the solutions used during the laboratory can be left on the laboratory bench to be picked up by laboratory personnel at the end of the period. Any volatile organic solvents should be placed in the hood in the container in which they were given to the student. Laboratory personnel will then dispose of these solvents.

Injuries

All injuries, no matter how minor, should be reported to the teaching assistant immediately.

Venipuncture and Processing of Blood Samples

During the first laboratory period, the student's blood is drawn by phlebotomists, and the blood samples are processed into plasma, serum, and whole blood cells. These blood products are used in the experiments described in this manual.

Blood Drawing

All blood donors must have fasted for 12 hours before giving blood to eliminate any effect of diet on the levels of the various components in the blood, i.e., lipids, glucose, electrolytes, and so on. Only water is allowed during this 12-hour fast. Sweet rolls, fruit juice, and coffee will be served after the blood has been drawn.

Preparation of Nuclei from Whole Blood

Pour 5 ml of whole blood (obtained from a 16 × 100 mm purple-topped tube, with an approximate draw of 10 ml, to which 15 mg dry EDTA-Na_2-has been added as an anticoagulant) into a 50 ml plastic conical centrifuge tube. Add 45 ml of Sucrose-Tris-Triton solution. Cap and mix gently by inversion. The tubes are centrifuged at 2,200 RPM (1,000 × g) for 10 minutes to obtain the nuclei. Decant the supernatant solution in one smooth motion, pausing briefly before returning the tube to the upright position. Add 10 ml of Sucrose-Tris-Triton solution to the tube and resuspend the pellet of nuclei by tapping the tube gently. Centrifuge the suspension at 2,200 RPM for 10 minutes and decant the supernatent solution as described above. Two hundred and thirty μl (use 100 μl and 10 μl pipettors) of Nuclear Lysis buffer are added to the cone area of the tubes in order to lyse the nuclei. The tubes are tapped gently until the pellets become detached from the bottom of the tubes. The lysed nuclei are frozen at −20°C (in the 50 ml tubes) for future use.

Preparation of the Serum
(whole blood minus blood cells, fibrinogen, and most of the clotting factors)

Approximately 10 ml of whole blood that has been collected in 16 × 100 mm red-topped tubes, with an approximate draw of 10 ml and with no additive, is allowed to clot. This takes 15–30 minutes. The clot is separated from the serum by centrifugation for 15 minutes at 2,200 RPM (1,000 × g). The serum is removed from the clot with a Pasteur pipette and placed in a flask or beaker. Aliquots of the serum are made and stored in the frozen state (−20°C) for the following determinations:

serum glucose	0.5 ml
serum thyroxine	0.5 ml
serum electrolytes	2.0 ml

Any remaining serum is frozen and stored.

Preparation of Plasma
(whole blood minus all blood cells)

Approximately 10 ml of the whole blood collected in purple-topped tubes containing an anticoagulant is centrifuged for 15 minutes at 2,200 RPM (1,000 × g). The plasma is separated from the cells with the aid of a Pasteur pipette and placed in a flask or beaker. Aliquots of plasma are made and stored in the frozen state for the following determinations:

plasma proteins	0.2 ml
plasma lipids	2.0 ml
plasma immunoelectrophoresis	0.2 ml

The aliquots of plasma for the lipid determinations are stored in Cryule™ vials at −70°C. Any remaining plasma is frozen and stored at −20°C.

Preparation of the Solution of Hemoglobin

The volume of the cells remaining in one of the tubes from which the plasma has been removed is estimated. The cells are washed several times with physiological saline (0.9% NaCl) in the following manner to remove the plasma proteins. A volume of physiological saline approximately equal to two times the volume of the cells is added to the cells, and the cells are gently mixed with the saline with a glass stirring rod. The suspension of the cells in saline is centrifuged for two to three minutes at 2,200 RPM (1,000 × g). The saline is removed from the cells with a Pasteur pipette. The suspension of the cells in saline and the centrifugation are repeated two more times to remove all plasma proteins from the red blood cells. After the final wash and centrifugation the red blood cells are hemolyzed (broken) by the addition of two volumes of H_2O and mixed with a stirring rod. The resulting red blood cell lysate (a solution of hemoglobin) is stored in the frozen state (−20°C) for use in the electrophoresis of hemoglobin and in the determination of glycosylated hemoglobin. Two 0.5 ml aliquots of the red blood cell lysate are needed for these procedures.

All samples should be labeled as follows:

Student Name ______________________

Group ______________________

Room ______________________

Day ______________________

Sample Code ______________________

Use this code for samples:

T = thyroxine

G = glucose

L = lipids

P = plasma protein electrophoresis

I = immunoelectrophoresis

HbG = glycosylated hemoglobin

D = DNA

X = extra plasma

H = hemoglobin electrophoresis

Y = extra serum

E = electrolytes

Medical Biochemistry

1 Electrophoresis of Blood Proteins

John F. Van Pilsum, Robert J. Roon, and Marilyn H. Koenst

Human blood contains hundreds of individual proteins. The quantitation of many of these proteins (or groups of proteins) in the plasma is used as a diagnostic aid by physicians. Electrophoresis makes possible the separation and quantitation of six groups of plasma proteins; immunoelectrophoresis permits the identification of 15 to 20 plasma proteins.

Electrophoresis is the movement of a charged particle or electrolyte in a solution under the influence of an electric field. The particles are in a solvent that is supported by an inert and homogenous stabilizing medium such as a paper or a gel. The movement of the proteins in the electrical field depends on their electrochemical and physical properties.

Immunoelectrophoresis involves separation of the proteins by electrophoresis in a gel followed by a diffusion of the proteins (at right angles to the migration of the proteins by electrophoresis) into an area into which antibodies to the proteins have been added. The combination of the serum proteins with their respective antibodies produces precipitin bands. This technique, which can be used to identify the serum proteins, is discussed in greater detail in Chapter 11.

Electrophoresis of serum or plasma proteins is a clinical procedure that is used by physicians in the diagnosis of a number of diseases, e.g., cirrhosis of the liver, protein malnutrition, nephrosis of the kidney, rheumatoid arthritis. If abnormal amounts of the γ-globulins are revealed by electrophoresis, the nature of the abnormality is further defined by the process of immunoelectrophoresis. For example, immunoelectrophoresis is used in the diagnosis of different types of myelomas (tumors of the bone marrow).

In today's experiments you will examine the patterns obtained after electrophoresis of your own plasma and hemoglobin. The migration of your hemoglobin will be compared with the migration of hemoglobins A, S, and C.

Principles

Factors Affecting the Net Charge on Protein Molecules

The main reason that proteins can be separated from one another by electrophoresis is that, with any given pH of the solution in which they are suspended, their net charges are not identical. The rate of migration of the particles of protein in the electric field depends mainly upon their charge.

Proteins are electrolytes whose net charge varies with the pH of the solution in which they are suspended. The pH of the solution at which the protein will not migrate in an electrical field is called the *isoelectric point*. A protein will not migrate at the isoelectric pt. because it has a net charge of zero. *Proteins have a net positive charge when suspended in solutions more acidic than their isoelectric point and have a net negative charge when suspended in solutions more alkaline than their isoelectric point. Negatively charged proteins (anions) will migrate toward the anode (+), and the positively charged proteins (cations) will migrate toward the cathode (−).*

Most proteins are electrolytes because a number of their amino-acid side chains have ionizable groups whose charges vary with the pH of the solution. All the ionizable groups of the amino acids are considered to be weak acids and can exist as the free acid HA (or HA^+) form, as the salt A^- (or A) form, or as

Table 1.1. The pK_a's of Ionizable Groups of Amino Acids in Proteins

Acid Type	Group	Acid Form	Salt Form	Location	pKa (Approx.)
Uncharged	Carboxyl	$C(=O)-OH$	$-C(=O)-O^-$	α-Carboxyl end-group*	3.5
	Carboxyl			Aspartyl- and glutamyl side-chains	4.0
	Phenol	$\geq C-OH$	$\geq C-O^-$	Tyrosine	10.0
	Sulfhydryl	$-SH$	$-S^-$	Cysteine	10.0
Cationic	Imidazolium	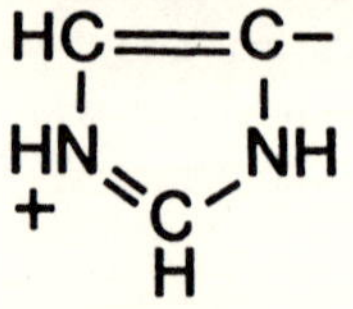 HC=C– / HN⁺–CH=NH (ring)	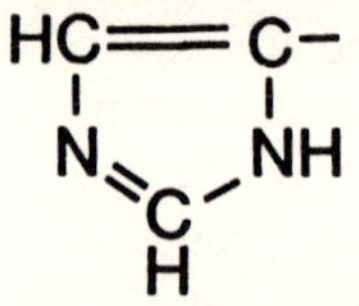 HC=C– / N=CH–NH (ring)	Histidine	7.0
	Ammonium	$-NH_3^+$	$-NH_2$	α-Amino* end-group	8.0
	Ammonium	$-NH_3^+$	$-NH_2$	ε-Amino of lysine	10.0
	Guanidinium	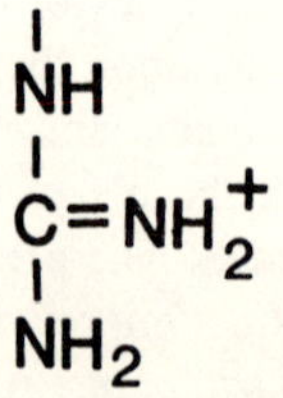 $-NH-C(=NH_2^+)-NH_2$	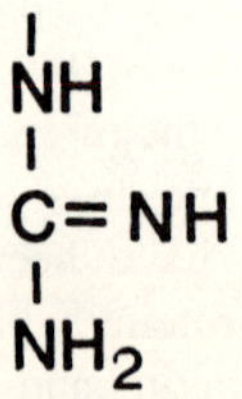 $-NH-C(=NH)-NH_2$	Arginine	12.5

*The pKa's of the α carboxyl groups and of the α amino groups of the free individual amino acids (i.e., not incorporated in proteins) are 2.4 and 9.6, respectively.

mixtures of the acid or salt forms. The acid groups of the type HA dissociate to yield a proton and the conjugate base of the weak acid, which has a negative charge. Groups of the type HA^+ dissociate to yield a proton and the conjugate base, which has no charge.

Note that for HA the acid and salt forms have a charge of 0 and −1, respectively; for HA^+ the acid and salt forms have a charge of +1 and 0, respectively.

$$1.\ HA \leftrightharpoons H^+ + A^- \qquad 1a.\ HA^+ \leftrightharpoons H^+ + A$$

The *dissociation constants* (K_a) of the amino-acid ionizable groups are defined by the equations shown below and have been experimentally determined.

$$2.\ K_a = \frac{[H^+][A^-]}{[HA]} \quad \text{or} \quad 2a.\ K_a = \frac{[H^+][A]}{[HA^+]}$$

The *pK_a* is defined as the negative log of the dissociation constant of the weak acid with either type of acid (1 or 1a).

$$pK_a = -\log K_a$$

This is similar to the definition of pH:

$$pH = -\log H^+$$

The extent to which the ionizable groups exist in the acid or salt form depends on the pK_a of the group and the pH of the solution. The pK_a's of the ionizable groups of the amino acids in proteins and the charge of the acid and salt forms are listed in Table 1.1. *For each ionizable group, the acid form predominates at pH values below its pK_a, and the salt form predominates at pH values above its pK_a.* The methods for calculating the exact charge on amino acid functional groups are outlined in the appendix to this chapter. Students should become familiar with these methods.

Factors Affecting Mobility of Proteins in an Electric Field

The mobility of the protein, that is, the direction and rate of migration in the electric field, is mainly dependent on the *net charge* of the protein. The *size* of the protein is also a factor in the rate of protein migration—small molecules move more rapidly than large molecules because they offer less resistance. *Shape* is also a factor; spherical proteins move with less resistance than elongated proteins of the same mass. The *viscosity* of the solvent determines solvent resistance to protein flow. The *supporting medium* should theoretically be inert, serving only to reduce convection currents (spreading of the bands of protein). In practice, however, the support medium can interact with the protein. Absorption or molecular sieving can retard the mobility of the protein, for example, in polyacrylamide or starch gels. Sieving is not a problem with agar gel because it has large pores.

If the support medium carries a charge, the phenomenon called *electroendosmosis* may greatly alter the mobility. Agar gels contain large numbers of negatively charged carboxyl and sulfate groups; therefore agar gels have a high negative charge. Agarose or cellulose acetate gels contain only negatively charged carboxyl groups; thus they have a smaller negative charge than agar gels. Polyacrylamide gels carry essentially no charged groups. Therefore electroendosmosis occurs to a greater extent in agar than in agarose or cellulose acetate and is not a factor in polyacrylamide gels. In negatively charged gels, the counter ions of the charged gels create a net solvent charge opposite to that of the gel. The electrophoresis of plasma proteins is done at alkaline pH (9.0). At this pH, all the proteins have a net negative charge and will migrate towards the anode (+). The positively charged component of the buffer, i.e., Na^+ ions (or the mobile counter ions) migrate to the cathode (-). As Na^+ ions migrate toward the cathode, they carry water molecules along. This stream of solvent is strong enough to move the negatively charged proteins towards the cathode—a direction opposite to that expected on the basis of the particle charge.

Characterization of Serum Proteins

Proteins differ greatly in their amino acid composition and therefore in their isoelectric points and in their mobility in an electric field. In electrophoresis of serum or plasma proteins, the discrete bands of individual proteins or groups of proteins can be quantitated by staining techniques. The isoelectric points (or ranges) and the molecular weights (or ranges) of serum proteins are listed in Table 1.2.

Table 1.2. Isoelectric Points and Molecular Weights of Human Serum Proteins

Protein	Isoelectric Point	Molecular Weight
Albumin	4.7	69,000
α_1-Globulins	1.8–4.1	41,000–195,000
α_2-Globulins	3.8–5.4	62,000–820,000
β-Globulins	5.3–5.9	85,000–20,000,000
γ-Globulins	6.3–7.3	150,000–900,000

In the electrophoresis of plasma or serum proteins, cellulose acetate strips provide the support medium, and the solvent is sodium barbital buffer, pH 8.6. Under these conditions, all the serum or plasma proteins have a net negative charge. However, because of their differences in isoelectric points, the magnitude of these net negative charges is *not* identical.

Patterns Obtained After Electrophoresis of Normal Human Plasma and Serum

The bands observed after the electrophoresis and staining of the proteins in normal human serum are shown in Figure 1.1. The proteins or groups of serum or plasma proteins that have been separated by electrophoresis can be quantitated by using a densitometer. The proteins on the support medium are stained with a protein dye and the support system is scanned in the densitometer. The optical densities of the protein stains and the areas of protein staining are used to calculate the amount of protein. The technique of densitometry is illustrated on page 7. Densitometry cannot be used in immunoelectrophoresis because of the complexity of the precipitin bands. The densitometer scan of the pattern obtained after electrophoresis of the serum is shown in Figure 1.1. This scan and the values depicted on Figure 1.1 are the data submitted to the physician by the clinical chemistry laboratory.

The densitometer scan in Figure 1.1 is interpreted as follows: The numbers recorded under the various protein peaks are called integration units. These are recorded by the densitometer and represent, in simple terms, the area encompassed under each peak, which is proportional to the amount of protein present. This sample had a total of 140 integration units, 77 of which are represented by albumin. The percent of albumin in the serum is therefore 77/140, or 55%. The complete analysis of the serum proteins is accomplished by having the value of the total amount of protein in the serum determined by an independent method such as the biuret reaction. The total serum protein in this case was 6.7 g/100 ml. Thus the grams of albumin per 100 ml for this sample were calculated to be 3.7.

The distribution of normal plasma proteins is shown in Table 1.3

Table 1.3. Distribution of Normal Human Plasma Protein Groups

Protein	(% of Total)
Albumin	55.2
α_1-Globulins	5.3
α_2-Globulins	8.7
β-Globulins	13.4
Fibrinogen	6.5
γ-Globulins	10.9

Clinical Applications

Diagnosis of Disease States Using Plasma Protein Electrophoresis

Electrophoresis as a diagnostic aid is illustrated in Table 1.4 and Figure 1.2.

Table 1.4. Normal Plasma Protein Concentration and Some Variations Found in Disease States

Protein	Normal Adult Range g/100 ml	Decrease	Increase
Albumin	3.5–5.4	Cirrhosis, malnutrition, nephrotic syndrome	Dehydration
α_1-Globulins	0.3–0.6	α_1-Antitrypsin deficiency	Acute and chronic infections
α_2-Globulins	0.4–0.9	Acute hepato-cellular necrosis	Rheumatoid arthritis
β-Globulins	0.6–1.1		Collagen disorders, chronic infections
Fibrinogen	0.2–0.4	Hypofibrino-genemia	Inflammation
γ-Globulins	0.7–1.5	Nephrotic syndrome Hypogamma-globulinemia	Multiple myeloma, chronic infections, collagen disorders

Occurrence of Hemoglobin Variants Hemoglobin Structure

Human hemoglobins have a molecular weight of 65,000 and are tetramers of four polypeptide chains, each with a molecular weight of ~17,000. Normal adult hemoglobins contain four related but electrophoretically distinct polypeptides, designated as α, β, γ, and δ. The following tetramers are found in normal human blood: $\alpha_2\beta_2$, $\alpha_2\gamma_2$, and $\alpha_2\delta_2$. The $\alpha_2\gamma_2$ form (fetal hemoglobin, called HbF) is predominant in the fetus. Just months prior to birth, the biosynthesis of γ-chains decreases and the synthesis of the β-chains increases. HbF comprises less than 0.5% of the total hemoglobin in the adult human, and $\alpha_2\beta_2$ (HbA) is the principal form present in the adult. The blood of normal adults also contains $\alpha_2\delta_2$ (HbA_2) at amounts equal to 2.5% of the total hemoglobin. The variations in the four normal human hemoglobin subunits (α, β, δ, and γ) are of genetic origin. The individual subunits of hemoglobin are coded for by four distinct genes.

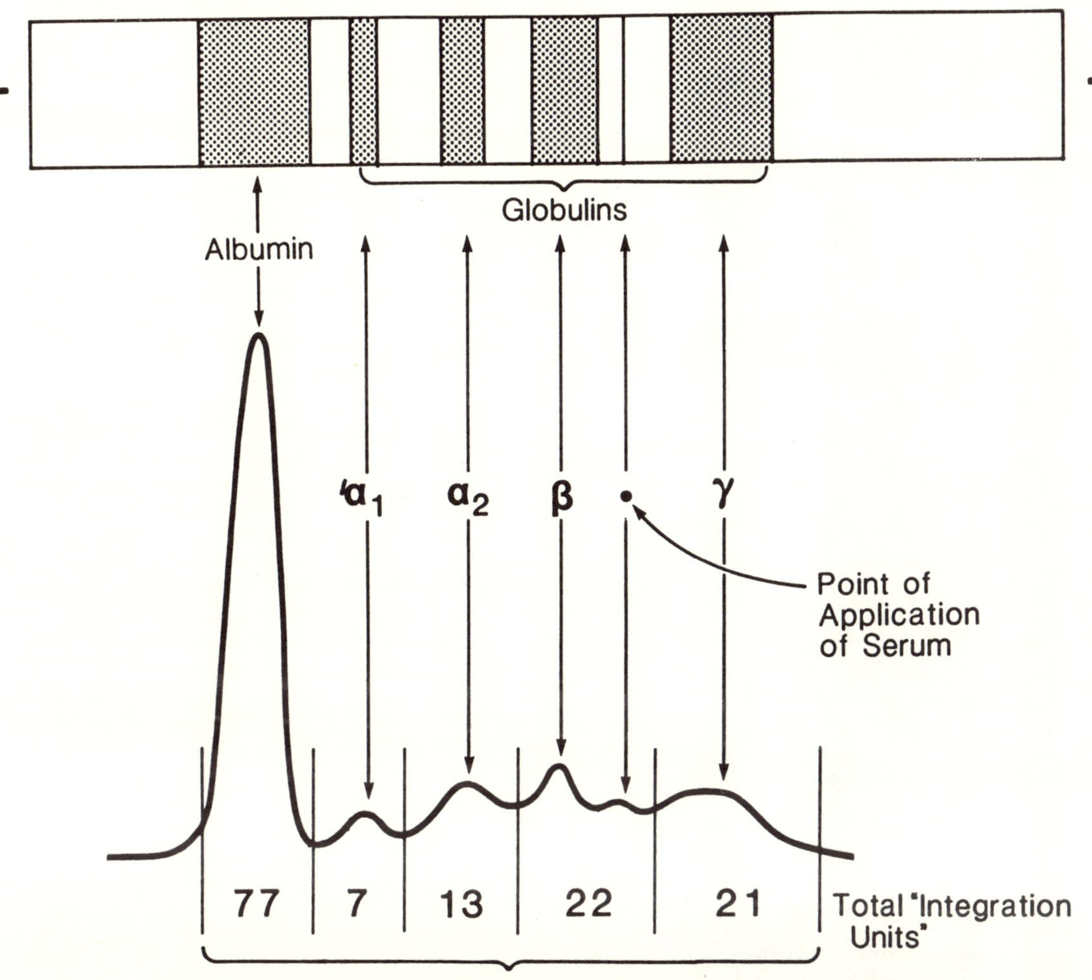

"Integration Units"
obtained with the densitometer

Total Serum Protein = 6.7 gm/100ml

	%	gm/100ml
Albumin	55	3.7
α_1-Globulins	5	0.3
α_2-Globulins	9	0.6
β -Globulins	16	1.1
γ -Globulins	15	1.0

Figure 1.1. Electrophoresis of Normal Human Serum

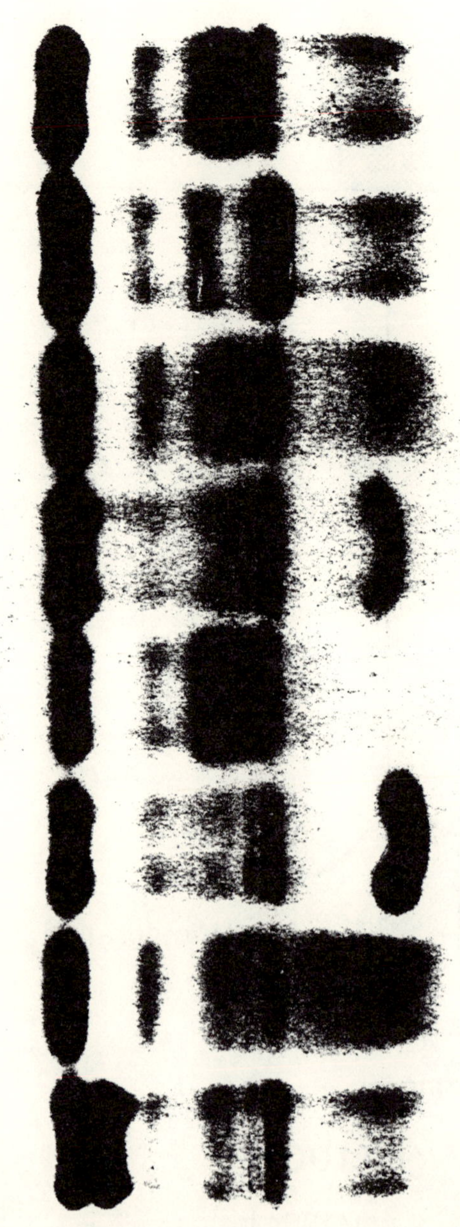

Figure 1.2. Electrophoresis patterns of normal human plasma and serum and some abnormal sera

Interpretation of Figure 1.2

1. Normal, or control, serum. The thin line in the β-region was the point of application of the serum for all samples. Electrophoresis was in barbital buffer, pH 8.6, using a Beckman Microzone system on cellulose acetate strips at 250 volts for 20 minutes. Under these conditions electroendosmosis is observed, that is, the β- and the γ-globulins appear on the cathode side of the application point.

2. Elevated β-globulins. This pattern is observed in iron-deficiency anemia, in which there is a stimulation of the synthesis of the iron-transport protein, *transferrin*. Large amounts of transferrin in serum are often found during the last months of pregnancy. The β-lipoproteins, (see Chapter 5), also migrate with the β-globulins. Patients with a severe Type II hyperlipoproteinemia would contain high amounts of proteins migrating with the β-globulins.

3. Normal or control, plasma. Plasma contains fibrinogen, whereas serum does not. The fibrinogen band is faintly apparent between the β- and γ-globulins.

4. Monoclonal protein in the γ-globulin. In normal serum or plasma the γ-globulins appear as a wide diffuse band because the proteins having γ mobility are a heterogeneous group of immunoglobulins. In disease states such as multiple myeloma and Waldenstrom's macroglobulinemia, a group of genetically-identical cells (a clone) will secrete a specific immunoglobulin. This is seen after electrophoresis as a discrete narrow band which is called a "monoclonal" peak. The monoclonal peak may be of β or γ mobility; this depends on the nature of the single protein produced by the clone of cells.

5. Nephrotic syndrome. This serum contains low amounts of albumin and γ-globulin and high levels of α_2-globulins. The small molecular weight proteins such as albumin cross the glomerular membrane and are excreted into the urine. The γ-globulins are hydrolyzed to small molecular weight proteins in the nephrotic kidney and are also excreted into the urine. The large amounts of α_2-globulins are the result of a stimulation of their synthesis in the liver in an attempt to compensate for the low osmotic pressure caused by the loss of albumin and γ-globulins from the blood.

6. Monoclonal protein in the γ region. Note the intense narrow band in the normally diffuse γ region. The total protein in this serum was 16g/dl as compared to a normal value of 6–8g/dl.

7. Hypergammaglobulinemia. This pattern indicates a "polyclonal" increase in the γ-globulins. This condition can be caused by the increased synthesis of normal heterogenous immunoglobulins (e.g., chronic infection). The increase in γ-globulins in cirrhosis is caused by the failure of liver cells to degrade the serum γ-globulins.

8. Bisalbuminemia. This rare pattern is not associated with any disease but is caused by a genetic variant. This variant albumin is seen most commonly in people of European descent and in the American Indian population.

Abnormal Hemoglobin Variants

A large number of mutations are found in human hemoglobin. These mutations cause alterations in the hemoglobin molecule. Most of the mutations result in the substitution of a single amino acid in the primary amino acid sequence of the globin portion of the hemoglobin. Some of these amino acid substitutions affect the function of the hemoglobin. The most common abnormal hemoglobin variant is sickle-cell hemoglobin (HbS). The sickle-cell mutation produces the substitution of a valine residue for glutamic acid in position 6 of the β-chain. Both normal (HbA) and sickle-cell (HbS) hemoglobin variants are found in a heterozygous individual. Sickle-cell hemoglobin (HbS) predominates in a homozygous person, along with significant amounts of HbF. HbC is a hemoglobin variant in which glutamic acid in position 6 of the β-chain is replaced by lysine. A great number of other abnormal hemoglobin molecules have been discovered—many of which do not exhibit abnormal functions. If the amino acid substitution in the peptide chain results in a change in the net charge of the

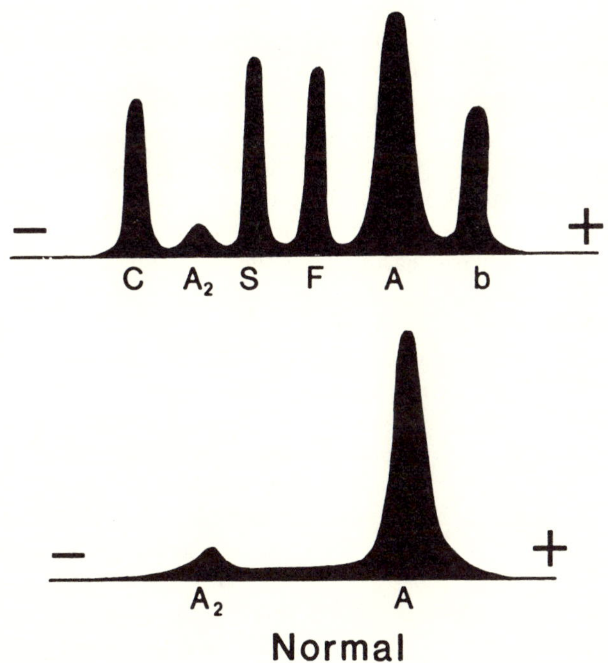

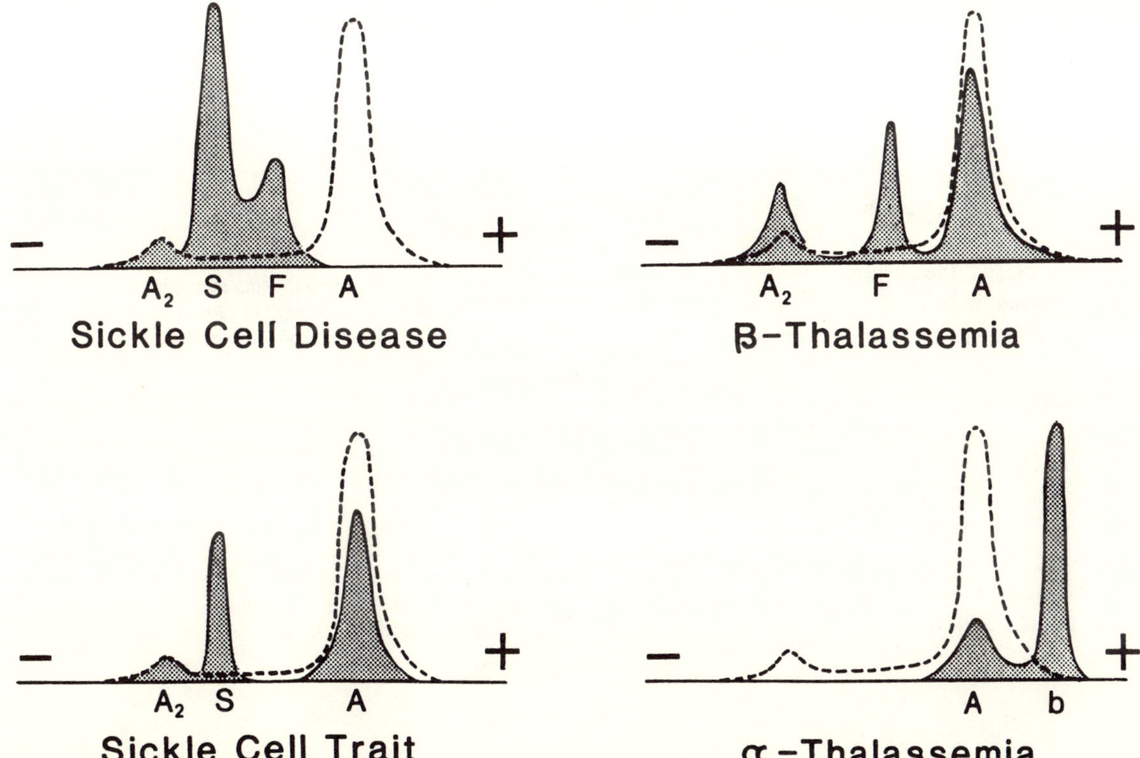

Figure 1.3. The relative mobility of some comon hemoglobin variants. A-predominant normal adult form; A_2-minor normal adult form; F-predominant normal fetal form; C-abnormal (amino acid substitution); S-abnormal (amono acid substitution); b-no α chains (β_4). reproduced from Gelman Instrument Company, Technical Bulletin No. 22, with their permission.

molecule, the hemoglobin will usually function abnormally. Depending on its location, an amino acid substitution (whether or not the charge is changed) may alter the solubility, subunit interaction, or ligand binding—all of which can change the physiological properties of the molecule.

Abnormal hemoglobin variants are also found in which the variation is in the amount of normal hemoglobin subunits, e.g., α- and β-Thalassemia. Individuals heterozygous for α-Thalassemia have a lower than normal capacity for the synthesis of α-chains. The hemoglobin of adults with this syndrome is predominately β_4. Adults with β-Thalassemia have a low capacity to synthesize β-chains, and their hemoglobin contains a large amount of HbF (fetal hemoglobin) and HbA_2 ($\alpha_2\delta_2$).

The relative mobility of common hemoglobin variants is illustrated in Figure 1.3.

Procedures

Plasma Proteins

1. The cellulose acetate strips are immersed in Na-barbital buffer, pH 8.6. A strip is placed on a saturation pad so that the notch in the strip is at the top and at the left. The strip is blotted with filter paper to remove excess buffer.

2. The strip is placed across the bridge in the electrophoresis chamber containing Na-barbital buffer so that the notch in the strip is at the top and the left. The strip should be immersed in buffer at each end only. If the portion of the strip on which the proteins migrate is immersed, the proteins will be removed from the strip by the buffer.

3. One or two drops of 0.1% bromphenol blue are added to your aliquot of plasma and mixed. A capillary tube is filled with the plasma-bromphenol blue mixture, and the sample is applied on the strip near its center. The sample should be applied evenly. During its application the ends of the strip should be held against the bridge to prevent the center of the strip from being immersed in the buffer.

4. The proteins are allowed to migrate at 225 volts for 60 minutes. Because voltage across the electrophoresis chambers is high enough to be a hazard, the teaching assistants will operate the electrophoresis apparatus.

5. The strips are removed from the chamber and placed in Ponceau S-fixative-dye solution for five minutes. The dye will combine with the proteins and thus indicate their location. Excess dye is removed from the strips by immersion in 5% acetic acid. The strips are allowed to dry.

Hemoglobin

1. The cellulose acetate strips are immersed in tris-EDTA-borate buffer, pH 9.2. A strip is placed on a saturation pad so that the notch in the strip is at the top and the left. The strip is blotted with filter paper to remove excess buffer.

2. The strip is placed in the chamber containing the tris-EDTA-borate buffer, and the red cell lysates are applied. Two hemoglobin samples are applied to each strip: the donor's own lysate on one half of the strip and a standard hemoglobin of a known type on the other half of the strip. The samples are applied on the cathode side of the strip.

3. The hemoglobins are allowed to migrate at 300 volts for about one hour.

4. The strips are removed from the chamber and allowed to dry. The relative rates of migration of the hemoglobins are determined.

Selected References

Fairbanks, V. F. 1976. Hemoglobin, Hemoglobin Derivatives, and Myoglobin. In *Fundamentals of Clinical Chemistry*, ed. N. W. Tietz, 2nd ed., pp. 401–454. Philadelphia, Saunders.

Grant, G. H. and Kachmar, J. F. 1976. The Proteins of Body Fluids. In *Fundamentals of Clinical Chemistry*, ed. N. W. Tietz, 2nd ed., pp. 298–376. Philadelphia, Saunders.

Nelson, D. A. 1979. Erythrocytic Disorders. In *Todd-Sanford-Davidsohn: Clinical Diagnosis and Management by Laboratory Methods*, ed. J. B. Henry, 16th ed., Vol. 1, pp. 964–1035. Philadelphia, Saunders.

Ritchie, R. F. 1979. Specific Proteins. In *Todd-Sanford-Davidsohn: Clinical Diagnosis and Management by Laboratory Methods*, ed. J.B. Henry, 16th ed., Vol. 1, pp. 228–258. Philadelphia, Saunders.

Problems

1. Draw the patterns obtained with the electrophoresis of your plasma. Designate cathode, anode, and point of sample application.

2. Explain the migration pattern of albumin, relative to the other plasma proteins, on the basis of its physical characteristics.

3. A mixture of three hemoglobins includes A, Q_{India} and $J_{Baltimore}$. In hemoglobin Q_{India}, histidine replaces aspartate at position 64 in the α-chains, and in hemoglobin $J_{Baltimore}$ aspartate replaces glycine at position 16 in the β-chains. The mixture is subjected to electrophoresis in cellulose acetate at pH 9.0. The isoelectric point of HbA is 6.8. Draw the expected pattern.

4. Draw the patterns obtained with the electrophoresis of normal and abnormal (S or C) hemoglobin.

5. $pK_a = 2.4$ $pK_a = 10.5$ $pK_a = 9.5$ $pK_a = 2.4$

```
       O     NH2   NH2 O
       ‖      |     |  ‖
HOCCH2CHCH2CHCOH
```

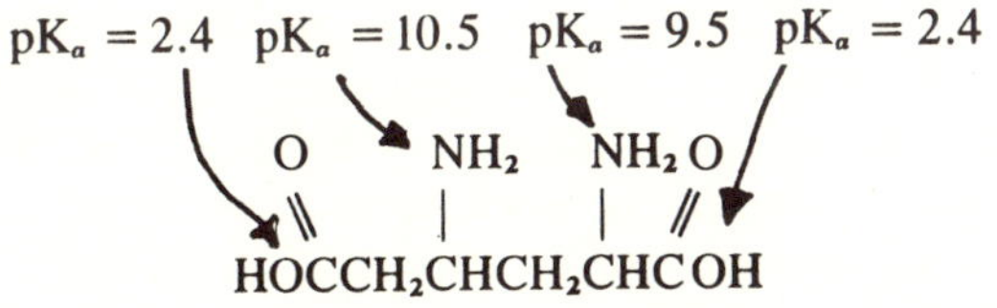

Calculate the net charge on the amino acid shown above at pH 1, 5, and 13.5. What net migration will occur if the amino acid is subjected to electrophoresis at each of these pH's?

6. If 10 ml of a 0.1 N solution of a weak acid, HA (pK_a = 4.5), is titrated with 1 N NaOH, what is the pH when 0.1 ml NaOH has been added? When 0.5 ml and 0.9 ml of NaOH have been added?

7. Three globular proteins of similar size and shape are subjected to electrophoresis at pH 10.0. If their pI's are 4, 6, and 8, respectively, what would be the expected order and direction of their migration?

8. In a genetic variant of hemoglobin, the amino acid aspartic acid is substituted for valine. What effect would this have on migration during electrophoresis conducted at pH 1? At pH 5? If lysine were substituted for valine, what would be the effect on migration at pH 1 and 5?

9. Even though plasma contains hundreds of different proteins, only a few bands are observed after electrophoresis. Why is this?

10. The clinical determination of nonprotein substances in blood (e.g., glucose, urea, creatinine) often requires the preparation of a protein-free filtrate of the blood. That is, the proteins are removed by precipitation with various reagents such as trichloroacetic acid, perchloric acid, or tungstic acid. Why will these reagents precipitate the blood proteins?

11. Orosomucoid is a glycoprotein in plasma that is not precipitated with the protein precipitants listed in problem 10. Orosomucoid has a molecular weight of 40,000, an isoelectric point of 2.7, and a carbohydrate content of 40%. Why will the protein precipitants *not* precipitate this protein?

Appendix

Calculation of the Charge on an Amino Acid

The effect of pH on the charge of a protein can be understood by examining the effect of pH on the charge of a simple amino acid, that is, the titration curve of an amino acid. The fact that proteins are buffers can also be explained by studying the titration curve. If a mole of the amino acid

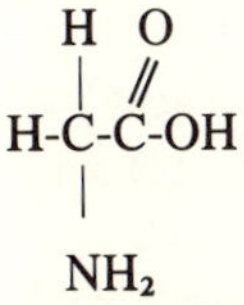

```
H O
| //
H-C-C-OH
|
NH2
```

(i.e., glycine) is dissolved in water and the pH of the solution adjusted to 1, the glycine has the configuration,

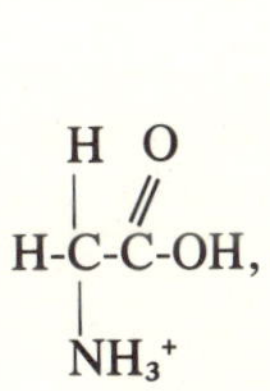

```
H O
| //
H-C-C-OH
|
NH3+
```

or, in other words, has a net charge of +1. The charge of the amino acid may be calculated for any pH by means of the Henderson-Hasselbalch equation for the dissociation of weak acids:

$$pH = pKa + \log \frac{salt}{acid}$$

(Henderson-Hasselbalch equation)

pH and pK_a are the negative logarithms of the hydrogen ion concentration and the dissociation constant of the ionizable groups, respectively. At pH 1, the charge of the amino group ($pK_a = 9.6$) is calculated as:

$$1 = 9.6 + \log\frac{\text{salt}}{\text{acid}}$$

$$\log\frac{\text{salt}}{\text{acid}} = -8.6$$

$$\frac{\text{salt}}{\text{acid}} = 10^{-8.6}$$

or

$$\frac{\text{salt}}{\text{acid}} \cong \frac{1}{1{,}000{,}000{,}000.}$$

Therefore at pH 1, the amino group exists as the acid (or NH_3^+) form, a charge of +1.

At pH 1, the charge of the carboxyl group ($pK_a = 2.4$) is calculated as:

$$1 = 2.4 + \log\frac{\text{salt}}{\text{acid}}$$

$$-1.4 = \log\frac{\text{salt}}{\text{acid}}$$

$$\frac{\text{salt}}{\text{acid}} = 10^{-1.4}$$

$$\text{or } \frac{\text{salt}}{\text{acid}} = \frac{1}{25}$$

Therefore the carboxyl group exists primarily as its acid (C(=O)-OH) form, a charge of 0. Thus the net charge of the amino acid at pH 1 is +1 +0 = +1.

The change in the charge of the amino acid with a change in pH and the ability of the ionizable groups of the amino acids to act as buffers become apparent upon examination of the titration curve for an amino acid. Alkali (NaOH) added to the solution of the amino acid glycine at pH 1 will first react with the most acidic group of the amino acid, the carboxyl group. The lower the pK_a of an ionizable group, the greater the tendency of the proton to dissociate—i.e., the greater its strength as an acid. For example, an acid that is 10% dissociated in a 1.0 N solution has a pK_a of 2.0.

$$\underset{(90\%)}{HA} \rightleftarrows \underset{(10\%)}{H^+} + \underset{(10\%)}{A^-}$$

$$K = \frac{[H^+][A^-]}{[HA]}$$

$$K = \frac{[0.1]\,[0.1]}{[0.9]}$$

$$K = \frac{0.01}{.9} \cong \frac{0.01}{1}$$

$$K = 0.01$$

$$\log \text{ of } 0.01 = -2$$

$$pK_a = 2$$

An acid that is 1% dissociated in a 1.0 N solution has a pK_a of 4.0.

$$K = \frac{[0.01]\,[0.01]}{[0.99]}$$

$$K = \frac{.0001}{.99} = \frac{0.0001}{1}$$

$$K = 0.0001$$

$$\log \text{ of } 0.0001 = -4$$

$$pK_a = 4$$

The titration curve obtained with an amino acid is shown in Figure 1.4. As the amount of alkali added approaches 0.5 equivalents per mole of amino acid, the rate of change in pH of the solution becomes minimal. In other words, a buffer system is being formed that consists of a mixture of the salt and acid form of the carboxyl group, i.e. $\frac{\text{C(=O)-O}^-}{\text{C(=O)-OH}}$. After the addition of 0.5 equivalents of NaOH per mole of amino acid, the salt-to-acid ratio of the carboxyl group equals 1, $\frac{0.5}{0.5}$, and the pH equals the pK_a (2.4) of the ionizable group.

$$pH = pK_a + \log\frac{0.5}{0.5}$$

$$pH = 2.4 + \log 1.0$$

$$\log 1.0 = 0$$

$$pH = 2.4 = pK_a$$

As the amount of alkali added approaches 1.0 equivalent per mole of amino acid, the rate of change in the pH of the solution increases and the buffering capacity is lost as the carboxyl group nears its complete conversion to the salt form. After the addition of 1.0 equivalent per mole of alkali, all of the carboxyl group is in the salt form, and the net charge of the amino acid is (−1) + (+1) or 0.

```
  H O
  | ||
H-C-C-O⁻
  |
  NH3+
```

The pH of the solution at which the amino acid has a net charge of 0 is called the isoelectric point (pI), and the amino acid is in its zwitterion form.

As alkali continues to be added, it next reacts with the proton of the amino group (a weaker acid than the

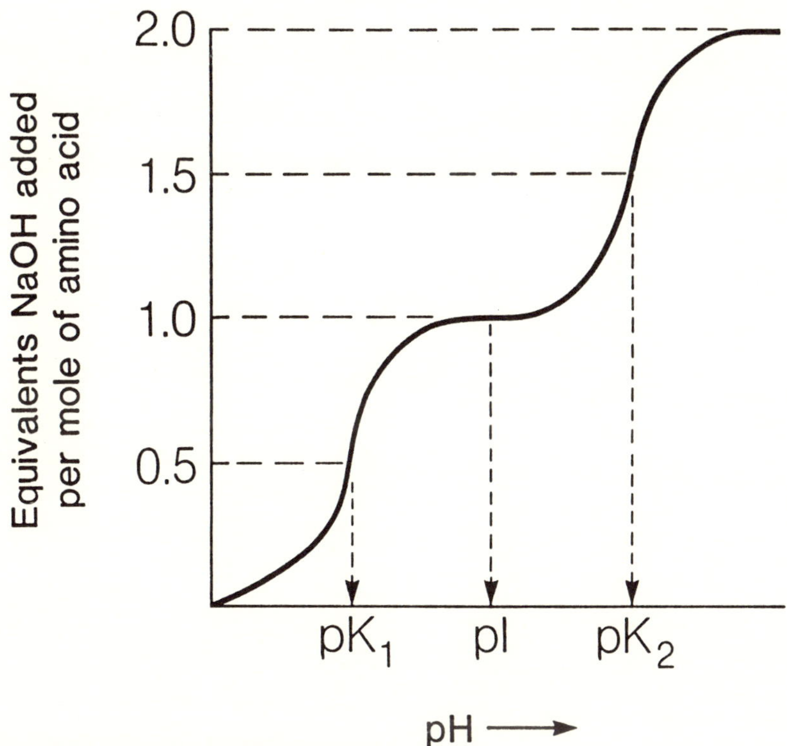

Figure 1.4. The titration curve of a mono-amino, mono-carboxylic amino acid

carboxyl group) and a second buffer system with a pK_a of ~9.6 is formed.

$$pH = pK_a + \log \frac{(-NH_2)}{(-NH_3^+)}$$

After the addition of 1.5 equivalents per mole of amino acid, the salt-to-acid ratio of the amino group equals 1, $\frac{0.5}{0.5}$, and the pH equals the pK_a of the ionizable group.

$$pH = pK_a + \log \frac{0.5}{0.5}$$

$$= 9.6 + \log 1.0$$

$$\log \text{ of } 1.0 = 0$$

$$pH = 9.6 = pK_a$$

Calculation of the Isoelectric Point of an Amino Acid

The isoelectric point for an amino acid with two ionizable groups may be calculated from the known pK_a's.

$$pI = \frac{pK_a \text{ of carboxyl group} + pK_a \text{ of amino group}}{2}$$

$$= \frac{2.4 + 9.6}{2} = 6$$

The isoelectric point of an amino acid with three ionizable groups may be estimated in a similar manner. It is first necessary to determine which two ionizable groups to use in the equation above. The two pK's to be used are the groups that dissociate on either side of the neutral species, i.e., the species with a net charge of 0. For example, the calculations of the pI for glutamic acid and lysine are as follows:

glutamic acid

```
O
//
C-OH←pK₂ = 4.3 (γ-carboxyl group)
|
HCH
|
HCH
|
HCNH₂←pK₃ = 9.7 (α-amino group)
|  O
| //
C-OH←pK₁ = 2.2 (α-carboxyl group)
```

The molecule can exist as four different charged forms.

```
         O               O          O          O
        //              //         //         //
       C-OH            C-OH       C-O⁻       C-O⁻
        |               |          |          |
       H-C-H     →     H-C-H  →   H-C-H  →   H-C-H
        |               |          |          |
       H-C-H     ←     H-C-H  ←   H-C-H  ←   H-C-H
        |               |          |          |
       H-C-NH3+        H-C-NH3+   H-C-NH3+   H-C-NH2
        |//O            |//O       |//O       |//O
       C-OH            C-O⁻       C-O⁻       C-O⁻
net
charge = (+1)          (0)        (−1)       (−2)
```

The two pK's to be used are those of the two carboxyl groups that dissociate on either side of the neutral species.

$$pI = \frac{pK_1 + pK_2}{2} = \frac{2.2 + 4.3}{2} = 3.25$$

```
        lysine
  NH2←pK3 = 10.5 (ε-amino group)
   |
 H-C-H
   |
 H-C-H
   |
 H-C-H
   |
 H-C-H
   |
 H-C-NH2←pK2 = 8.95 (α-amino group)
   |//O
   C-OH←pK1 = 2.2 (α-carboxyl group)
```

The molecule can exist in four different charged forms.

```
          NH3+          NH3+         NH3+        NH2
           |             |            |           |
         (CH2)4   ⇄    (CH2)4   ⇄   (CH2)4  ⇄   (CH2)4
           |             |            |           |
        H-C-NH3+      H-C-NH3+      H-CNH2      H-CNH2
           |//O          |//O         |//O        |//O
          C-OH          C-O⁻         C-O⁻        C-O⁻
net
charge =  (+2)          (+1)          (0)         (−1)
```

The pK's of the groups that dissociate on either side of the neutral species are those of the α- and the ε amino groups.

$$pI = \frac{pK_2 + pK_3}{2} = \frac{8.95 + 10.5}{2} = 9.7$$

Calculation of the Charge on a Peptide

As with the individual amino acids, the charges of peptides and proteins vary with the pH of the solution in which they are suspended. The isoelectric point for a protein or peptide is the pH of the solution at which the protein or peptide has a net charge of 0. The charge of a protein or peptide may be calculated at any given pH by using the Henderson-Hasselbalch equation. For example, the net charge of the peptide histidylaspartyllysylglutamylarginine is −1 at pH 9.0.

```
                                                              NH2+
  HC-NH                                                        ||
  ||   \CH                                                     C— NH2
  ||   //                                                      |
  C-N        COO⁻        NH3+         COO⁻                     NH
  |           |           |            |                       |
  CH2        CH2        (CH2)4        (CH2)                  (CH2)3
  |    O      |    O      |    O       |     O                 |    O
  |   //      |   //      |   //       |    //                 |   //
H2N-CH-C-NH-CH-C-NH-CH-C-NH-CH-C-NH-CH-C-O⁻
  histydyl   aspartyl     lysyl       glutamyl       arginine
```

The net charge of the amino group (histidyl α-amino group) at pH 9 is 0.

$$9 = 8.0 + \log \frac{\text{salt}}{\text{acid}}$$

$$\log \frac{\text{salt}}{\text{acid}} = 1.0$$

$$\log 10 = 1.0$$

$$\frac{\text{salt}}{\text{acid}} \text{ ratio} = \frac{10}{1}, \text{ or } \frac{10\ NH_2}{1\ NH_3^+}$$

The charge of the imidazole group (histidyl imidazole group) will also be 0, by the same reasoning.

$$9.0 = 6.0 + \log \frac{\text{salt}}{\text{acid}}$$

$$\log \frac{\text{salt}}{\text{acid}} = 3$$

$$(\log 1000 = 3)$$

$$\frac{\text{salt}}{\text{acid}} \text{ ratio} = \frac{1000}{1}$$

The charge of the carboxyl group in the aspartic acid side chain will be -1.

$$9.0 = 4.0 + \log \frac{\text{salt}}{\text{acid}}$$

$$\log \frac{\text{salt}}{\text{acid}} = 5$$

$$(\log 100{,}000 = 5)$$

$$\frac{\text{salt}}{\text{acid}} \text{ ratio} = \frac{100{,}000}{1}$$

The charge of the amino group on the lysine side chain will be essentially $+1$.

$$9.0 = 10 + \log \frac{\text{salt}}{\text{acid}}$$

$$\log \frac{\text{salt}}{\text{acid}} = -1$$

$$(\log 0.1 = -1)$$

$$\frac{\text{salt}}{\text{acid}} \text{ ratio} = \frac{1}{10}$$

By similar calculations, at pH 9.0 the charge of the carboxyl group on the glutamic acid is -1, the guanido group of arginine is $+1$, and the terminal carboxyl group is -1. Thus, the net charge of the peptide at pH 9 is:

$$0 + 0 + (-1) + 1 + (-1) + 1 + (-1) = -1.$$

2 Enzymes As Diagnostic Indicators

John D. Lipscomb and James B. Howard

The clinical application of enzymology was made possible by advances over the past 50 years in our understanding of cellular metabolism and enzyme kinetics. Enzymes are currently used in the clinical laboratory in three ways:

1. *They are reagents* employed in specific methods for measuring the concentrations of metabolites, such as glucose, in the blood.
2. The measurement of the *activity of enzymes* in the blood can serve in the diagnosis and management of certain diseases.
3. The *characteristics of the enzymes* in the blood can be used to identify the organ or tissue involved in the disease state.

During the next two weeks you will perform each of the three types of enzymatic assays. You will be working with one particular enzyme of great clinical importance: *lactate dehydrogenase (LDH)*. You will:

1. use a purified preparation of LDH to determine the amount of pyruvate, a metabolite in human blood;
2. determine the activity of endogenous LDH in a sample of human blood; and
3. do an electrophoretic separation of LDH taken from different organs of the rat—this will demonstrate that the isoenzyme distribution of LDH varies in different organs, a fact that can be used to pinpoint the location of a diseased tissue.

In a clinical laboratory enzyme assays are routine procedures for the laboratory technician. In this laboratory you will be designing an enzyme assay. This should give you an understanding of *how* and *why* these assays function. The major objective of these laboratory sessions is to help you comprehend the principles underlying enzyme assays.

Principles

General Principles of Enzyme Assay

The enzymatic assay is based on determining the rate at which the enzyme catalyzes the conversion of a metabolite (substrate) to a chemically altered form (product). Very few clinical assays monitor the metabolic substrate or product concentrations directly, however, because of difficulties in quantitating these substances. Also, since the enzymatic assays are fundamentally *kinetic* procedures, the reaction rates must be determined at various times. Therefore the method of detection must be simple and rapid. Fortunately many enzymes require the presence of a small molecule called a coenzyme for the reaction to occur, and, in many cases, the chemical changes that occur in the coenzyme are much easier to detect than the changes in the metabolite itself. For instance, several coenzymes change color during the reaction owing to oxidation/reduction or other chemical processes. Therefore a spectrophotometer can be used to quickly quantitate the amount of coenzyme that has undergone reaction. This concentration is directly correlated with the amount of metabolite that has been converted to product.

A large number of enzymes, including lactate dehydrogenase (LDH) and many other dehydrogenases, require the coenzyme nicotinamide adenine dinucleotide (NAD^+) or its phosphorylated analog ($NADP^+$) for their catalytic activity. NAD^+ can also exist in a more reduced form (NADH). The oxidized and reduced forms of nicotinamide adenine dinucleotide are shown in Figure 2.1.

Note that the oxidation of NADH to NAD^+ involves

the loss of a hydride ion, i.e., two electrons and one proton. A second proton is gained by the reduced substrate from water to maintain charge stoichiometry. LDH catalyzes the oxidation of NADH to NAD^+, and there is a concomitant reduction of pyruvate to lactate. LDH also catalyzes the reverse reaction in which the energy released by lactate oxidation is used to reduce NAD^+ to NADH.

$$CH_3\text{-}\overset{O}{\overset{\|}{C}}\text{-}\overset{O}{\overset{/\!/}{C}}\text{-}O^- + NADH + H^+ \rightleftharpoons CH_3\text{-}\underset{H}{\underset{|}{\overset{OH}{\overset{|}{C}}}}\text{-}\overset{O}{\overset{/\!/}{C}}\text{-}O^- + NAD^+$$

pyruvate lactate

The assay of enzymes that require NAD^+ is facilitated by the fact that NAD^+ and NADH have different UV-visible absorption spectra (Figure 2.2).

Note that NADH absorbs light at 340 nm, whereas NAD^+ does not. Therefore the reaction in either the forward or reverse direction can be measured spectrophotometrically by observing the absorbance change at 340 nm. In the formation of lactate from pyruvate, the absorbance will decrease as the NADH is utilized. In the formation of pyruvate from lactate, the absorbance will increase as NADH is produced. The following should also be apparent: one mole of NADH is formed for every mole of lactate that is converted to pyruvate. And conversely one mole of NADH is utilized for every mole of pyruvate converted to lactate. The equilibrium constant for LDH is approximately 10,000 in favor of lactate formation. Since this reaction has two substrates (lactate or pyruvate and NAD^+ or NADH), the complete kinetic analysis is complex. Fortunately we can assume simple, single substrate Michaelis-Menten kinetics if one of the substrates is always at saturating concentrations, e.g., 25–50 times its Michaelis-Menten constant (K_m), as NADH is in this case. (The Michaelis-Menten constant is the concentration of the substrate that allows the enzyme to proceed at one-half of its maximal velocity.) This means that we can analyze the kinetics with the use of the following Michaelis-Menten equation:

$$v_0 = \frac{V_{max}\,[S]}{K_m + [S]} = \frac{TON[E_T][S]}{K_m + [S]}\,.$$

TON = turnover number = μmoles product formed per μmole enzyme active sites per minute, at saturation

S = substrate concentration

V_{max} = turnover number × total enzyme active site concentration

Oxidized Coenzyme

NAD^+

nicotinamide adenine dinucleotide

hydride ion

Reduced Coenzyme

NADH

reduced nicotinamide adenine dinucleotide

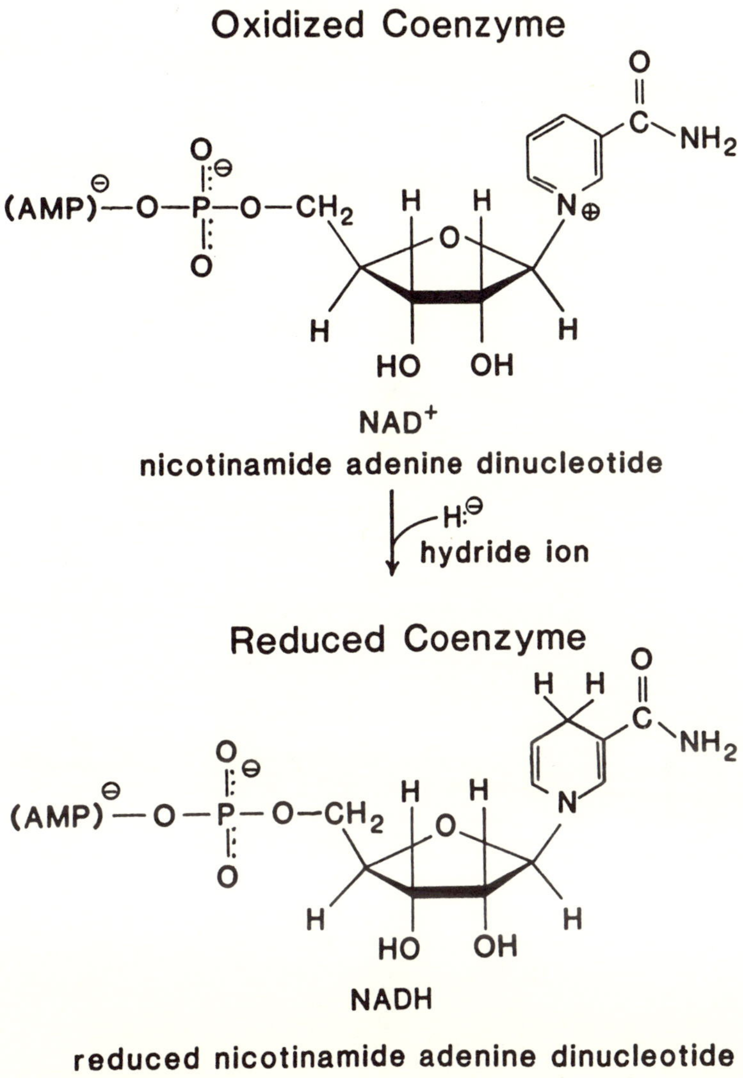

Figure 2.1. The oxidized and reduced forms of nicotinamide adenine dinucleotide.

E_T = enzyme concentration (active site concentration as used here)

v_0 = initial velocity

K_m = Michaelis-Menten constant

Determination of K_m and Turnover Numbers of an Enzyme

Two quantities can be ascertained by measuring the initial velocity of the enzyme reaction (v_o). First we can determine the *substrate concentration*—if we know the K_m, the enzyme turnover number (TON), and the enzyme concentration (E_T). Second we can find the *enzyme concentration*—if we know the K_m, enzyme turnover number, and substrate concentration. It should be apparent that two intrinsic

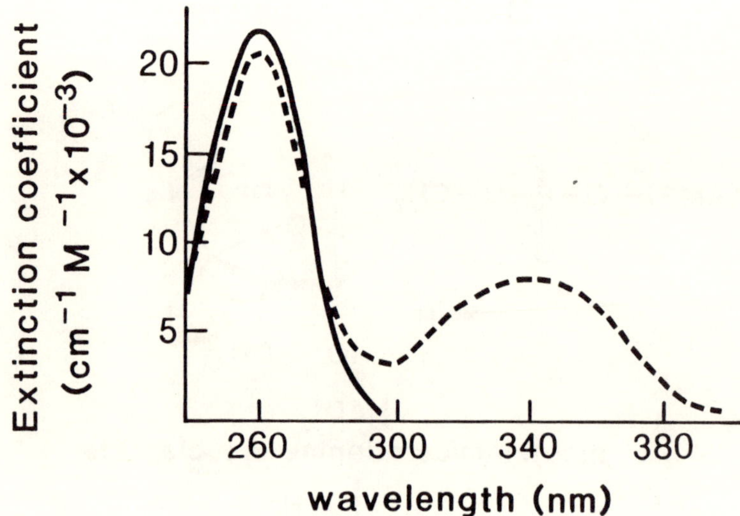

Figure 2.2. Absorption spectra of NAD^+ (———) and NADH (----). The extinction coefficient of NADH at 340 nm is 6.22 x 10^3 $cm^{-1}M^{-1}$. The absorption at 260 nm in both cases is due to the adenine moiety of the coenzyme.

properties of an enzyme must be known, i.e., K_m and turnover number, before the enzyme can be used as an analytical reagent.

The initial velocity for the LDH reaction can be determined by recording the change in absorbance at 340 nm at various time intervals and plotting the results, obtained with different concentrations of pyruvate (x, y, and z), as shown in Figure 2.3. The concentration of NADH at any time during the reaction can be calculated from the absorbance determinations at 340 nm by the equation

$$A = E_M \times L \times C \text{ (Beer's Law)},$$

where *A* is the *absorbance*, E_M is the *molar extinction coefficient* (6.2×10^3 cm^{-1} M^{-1} for NADH at 340 nm), *L* is the *path length of the reaction cell*, and *C* is the *concentration in moles liters*$^{-1}$. Thus the time course of the enzyme reaction can be monitored by absorbance changes at 340 nm.

The initial velocity for each of the enzyme reactions shown in Figure 2.3 will be equal to the slope of a tangent to the initial (linear) portion of its time-course curve. If the initial velocities for different initial substrate concentrations (i.e. data in Figure 2.3) are plotted against the initial substrate concentration, a hyperbolic curve is formed (See Figure 2.4).

In order to get more accurate values for K_m and V_{max}, the data from Figure 2.4 should be plotted in the double reciprocal form. This is a Lineweaver-Burk plot (see Figure 2.5).

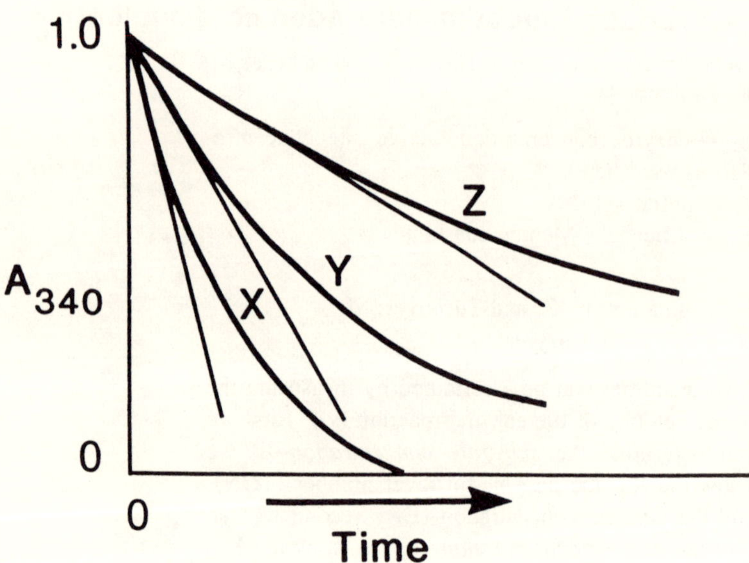

Figure 2.3. Determination of initial velocities of the LDH reactions at different pyruvate concentrations.

Experimental Conditions for Measuring Unknown Substrate

At low substrate levels, the K_m is much greater than the concentration of the substrate. Thus $K_m \gg [S]$, [S] can be ignored in the denominator and the Michaelis-Menten equation, $v_0 = \frac{V_{max}[S]}{K_m + [S]}$, reduces to $\frac{V_{max}[S]}{K_m}$, but V_{max} and K_m are constants at fixed enzyme levels, so $v_0 = M[S]$, where M is a constant equal to V_{max}/K_m. Thus v_o is proportional to [S].

Thus, at low substrate levels, the observed initial velocity is approximately a linear function of the substrate concentration. In enzyme assays designed to measure unknown substrate (metabolite) levels, it is convenient to use substrate concentrations that fall within this "linear" response region.

As substrate levels are increased above this region, it becomes progressively more difficult to accurately estimate the substrate levels because there is a smaller and smaller change in v_o (initial velocity) for a given incremental increase of substrate. Nevertheless it is still possible to determine the concentration of unknown substrate in the region where [S] is *not* small compared to K_m. To do this, however, the complete, unsimplified Michaelis-Menten equation must be used, because [S] is no longer proportional to v_o. (The details of this method will be discussed further in the sample calculation at the end of this section.) Note that when $[S] \gg K_m$, it becomes experimentally impossible to determine an unknown substrate concentration because $v_o = V_{max}$, a constant.

Experimental Conditions for Measuring the Amount of an Enzyme

At high substrate levels, $[S] \gg K_m$. K_m can be ignored and the equation $v_0 = \frac{V_{max}[S]}{K_m + [S]}$ reduces to $\frac{V_{max}[S]}{[S]} = V_{max} = \text{TON}\,[E_T]$ or, $\frac{v_0}{\text{TON}} = [E_T]$. *Thus at high substrate levels the observed initial velocity at a fixed enzyme* level is essentially a constant $(= V_{max})$, since the reaction is occurring at its maximum velocity, i.e., the enzyme is saturated and v_o is more or less independent of substrate concentration although it is *linearly dependent on enzyme concentration*. Therefore in assays designed to determine the amount of enzyme present in a sample, it is helpful to use high substrate levels for the following reasons:

1. The measured v_o is not sensitive to small deviations in the initial substrate concentration.
2. The v_o will be maintained for a longer period of time, making it easier to measure experimentally.
3. The relatively large amount of product obtained will facilitate measurement of v_o.

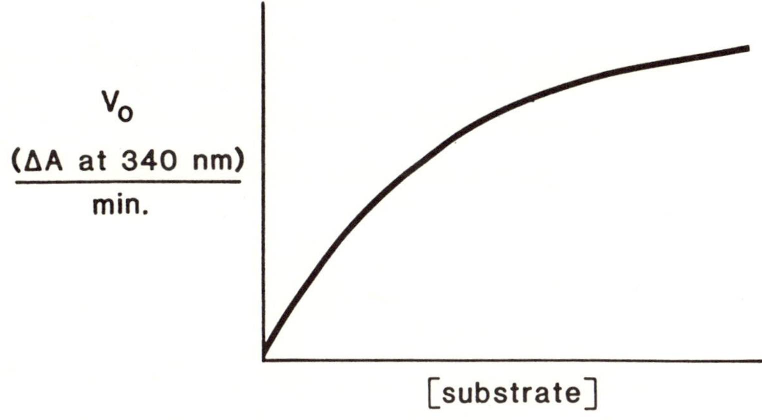

Figure 2.4. The initial velocity of an enzyme reaction as a function of its substrate concentration.

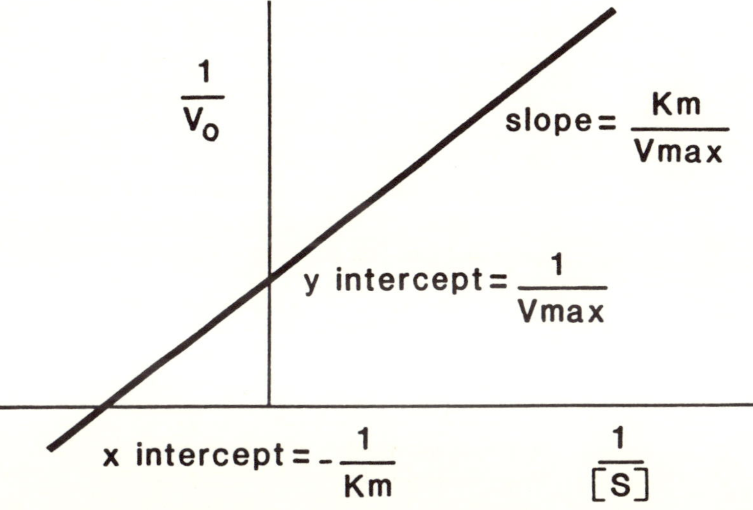

Figure 2.5. Lineweaver-Burk plot of the initial velocity of an enzyme reaction as a function of its substrate concentration

Clinical Applications

Measurement of Metabolites in Human Blood

The measurement of the concentration of metabolites in human blood and urine is of major diagnostic importance because disease processes often produce abnormally large and characteristic changes in the concentration of these substances. For example, excessive amounts of glucose are found in the blood and urine of patients with diabetes. The amounts of most metabolites can be determined by chemical (i.e., colorimetric) procedures. Unfortunately, colorimetric procedures are frequently not specific since they cannot distinguish between different metabolites having chemically similar functional groups. That is, in a complex fluid such as serum, the colorimetric reagents can react with several substances in addition to the one being measured. This problem has been alleviated by the development of assays that exploit the specificity inherent in enzyme catalyzed reactions. Most enzymes can select their single, metabolite substrate from complex mixtures and catalyze a specific chemical change in its structure. The rate of this reaction can be used to determine the concentration of the metabolite in the fluid. Examples of metabolites that are quantified in the clinical laboratory by using the technique of enzymatic assays are given in Table 2.1.

Measurement of Enzyme Activities in Human Blood

The measurement of enzyme levels in humans is important for two reasons: some enzymes are released into the blood in elevated levels in certain diseases and

Table 2.1. Metabolites of Clinical Importance that Are Assayed Enzymatically

Metabolite	Related Disease	Enzyme Used in the Assay
Fructose	Fructosuria	Fructose kinase
Galactose	Galactosemia	Galactose oxidase
Glucose	Diabetes	Glucose oxidase
Lactate or pyruvate	Liver disease	Lactate dehydrogenase
Urea	Kidney disease	Urease
Uric acid	Gout	Uricase

some diseases are the result of low enzyme activities in blood or tissues.

Advances in clinical biochemistry have permitted a new approach to diagnosing diseases: use of isolated, purified metabolites to detect the concentration of the enzymes that mediate their metabolism in blood and urine. Thus not only are enzymes being used to accurately measure substances in biological fluids, but they themselves can be quantitated and used as extremely valuable diagnostic indicators. Examples of clinical application of this type of enzyme assay are given in Tables 2.2 and 2.3.

Table 2.2. Enzyme Activities Quantified in the Laboratory to Indicate a Disease

Enzyme in Blood Showing Increased Level	Diseased Tissue Indicated
Acid phosphatase	Prostate
Alcohol dehydrogenase	Liver
Creatine kinase	Skeletal and cardiac muscle
Glutamate-oxaloacetate transaminase	Heart, liver
Glutamate-pyruvate transaminase	Liver
Isocitrate dehydrogenase	Liver, prostate
Lactate dehydrogenase isozymes	Liver, heart, or kidney
Lysozyme	Kidney

Table 2.3. Examples of Diseases Caused by Low Levels of Enzyme Activities in Tissues

Defective Enzyme	Resulting Disease
Arylsulfatase A	Metachromatic Leucodystrophy
Galactose-1-phosphate uridyl transferase	Galactosemia
Glucocerobrosidase	Fabry's Disease
Glycolytic enzymes in the erythrocyte	Hemolytic Anemia
Homogentisic acid oxidase	Alcaptonuria
α-keto acid decarboxylase	Maple Syrup Disease
Phenylalanine hydroxylase	Phenylketonuria (PKU)

Procedures

Two weeks are devoted to the experiments on lactate dehydrogenase. During the first week each group of four students calculates the K_m and turnover number of LDH to determine the concentration of an unknown solution of pyruvate and of LDH. During the second week the experiments on *isozymes* of LDH are performed.

Determination of K_m and Turnover Number for LDH with Pyruvate as the Substrate

To determine the K_m and turnover for LDH with pyruvate, assays are run under the following protocol: 1) Zero the spectrophotometer with buffer (the teaching assistant will provide instruction for the instrument at your lab bench); 2) into a 3 ml spectrophotometer cuvette add 2.5 ml of buffer, 0.1 ml of NADH (4.8 mM), and 0.4 ml of sodium pyruvate solution; 3) place in the spectrophotometer and measure the absorbance at 340 nm (this reading is a check on the pipetting procedures and should be the same for each assay); 4) add 10 μl of enzyme solution to the cuvette. Mix assay solution by placing a piece of parafilm over the cuvette and inverting several times. Put the cuvette into the spectrophotometer, record the absorbance, and simultaneously start the timer. Then read the absorbance at the shortest possible intervals (this varies with the instrument at your station) until the initial velocity is clearly established and the reaction begins to slow down (one to three minutes). Repeat the assay for each of the different concentrations of Na pyruvate provided at your work area. *Be sure to record the concentration of pyruvate and enzyme given on the stock solutions.* The NADH and enzyme should be kept on ice. The buffer and pyruvate should be kept at room temperature.

Include the following calculations for your report.

1. *Enzyme concentration in molarity.* Record the absorbance of pure enzyme solution at 280 nm. (This number will be given to you). The E_M for beef heart LDH is $2.09 \times 10^5 (\text{cm M})^{-1}$ at 280 nm. If there are four subunits per molecule, what is the "active site" concentration in the assay? (See p. 22)

2. *Lineweaver-Burk Plot.* Plot the absorbance at 340 nm against time for your five assays. Considering the manual mixing and data collection techniques used in the experiment, it is best to use the first data point recorded after starting the reaction (by the addition of the enzyme) as the 0 time point. Determine the initial velocity from this plot for each initial substrate con-

centration and make a Lineweaver-Burk plot. (You will need to convert absorbance change into concentration by using Beer's Law.)

3. K_m and V_{max}.
4. The turnover number for LDH.

Determination of Pyruvate in an Unknown Solution

Obtain an unknown pyruvate solution from a teaching assistant. Use the assay method described above to determine the initial velocity caused by the unknown (undiluted) pyruvate solution. The assay is repeated on solutions of the unknown pyruvate that have been diluted 1:3 (i.e., 1 volume pyruvate + 2 volumes of buffer) and also diluted 1:10 (i.e., 1 volume of pyruvate + 9 volumes of buffer). This is to ensure that an appropriate concentration of pyruvate is used for *its* accurate quantitation—that is, the pyruvate must be present at a concentration that does not completely saturate the enzyme (see p. 18).

Calculate the concentration of pyruvate in the unknown. (Include all calculations in your report.)

Determination of LDH in an Unknown Solution

Obtain an unknown LDH solution from a teaching assistant. Use the assay method described above to determine the initial velocity caused by the unknown enzyme solution with the 5 mM pyruvate solution.

Calculate the concentration of enzyme in the unknown. (Show all calculations.)

Sample Calculations

Example of Calculation of K_m and V_{max} for LDH

The absorbance at 280 nm of an LDH stock solution was found to be 1.045. This absorbance at 280 nm is used to calculate the concentration of the LDH in the stock solution (see p. 22).

Since the initial velocity (v_o) of the LDH enzyme reaction is directly proportional to the observed change in absorbance (A) per unit time, the absorbance at 340 nm against time is plotted for each of the five concentrations of pyruvate used. Lines that originate at the initial A point are drawn tangent to each of the five curves and then the slopes of the tangent lines are calculated. These slopes (ΔA/min) are directly proportional to the initial velocity (v_o) of the reactions.

For example, assume the following data obtained are those in the accompanying tabulation.

Concentrations of the pyruvate stock solutions	*Initial ΔA_{340}/min (slope)*
0.375 mM	0.30
0.750 mM	0.57
2.25 mM	1.43
7.5 mM	3.11
22.5 mM	4.67

The ΔA/min measurements are converted into concentration changes of NADH in order to obtain the v_o values by dividing the ΔA/min by 6.2×10^3 (cm M)$^{-1}$, which is the extinction coefficient for NADH.

ΔA/min (slopes)	*Moles per liter/min = v_o*[a]	*$\frac{1}{v_o}$ min/moles per liter*
0.30	4.8×10^{-5}	0.21×10^5
0.57	9.1×10^{-5}	0.11×10^5
1.43	23×10^{-5}	0.043×10^5
3.11	50×10^{-5}	0.021×10^5
4.67	75×10^{-5}	0.013×10^5

[a]If the units of v_o concern you, see the note at the end of the example.

Note that the pyruvate in the incubation mixtures is less concentrated than it is in the original stock solution because 0.4 ml of the stock solutions were diluted to a final volume of 3 ml by the addition of the rest of the components (NADH, LDH, and buffer). Thus the concentrations of the pyruvate in the incubation mixtures are obtained by dividing the concentrations of each of the stock solutions by 7.5.

Stock pyruvate	*Assay conc. = S*	*$\frac{1}{S}$*
0.375 mM	0.05 mM	2.0×10^4 M^{-1}
0.750 mM	0.10 mM	1.0×10^4 M^{-1}
2.25 mM	0.30 mM	0.33×10^4 M^{-1}
7.50 mM	1.0 mM	0.10×10^4 M^{-1}
22.50 mM	3.0 mM	0.01×10^4 M^{-1}

Plot $\frac{1}{v_0}$ against $\frac{1}{S}$, draw the best straight line and read the slope, y intercept, and x intercept. Make the following calculations:

$$V_{max} = \frac{1}{\text{y int}} = \frac{1}{1 \times 10^3 \text{ min/M}}$$

$$= 1 \times 10^{-3} \text{ moles per liter/min}$$

$$K_m = -\frac{1}{\text{x int}} = \frac{1}{0.1 \times 10^4 \text{ M}} = 1.0 \times 10^{-3} \text{ M}$$

or K_m = Slope × V_{max} = (1.0 min) × (1×10^{-3} moles per liter/min) = 1.0×10^{-3} M. Calculate the enzyme concentration in the stock and correct for dilution into the assay mixture. The absorbance of the enzyme solution is 1.045 in this case, and 10 μl is diluted to 3.0 ml in assay.

$$\text{Conc. enzyme} = \frac{1.045\ \text{cm}^{-1}}{2.09 \times 10^{5}\ \text{M}^{-1}\ \text{cm}^{-1}}$$

$$= 0.5 \times 10^{-5}\ \text{M}$$

$$0.5 \times 10^{-5}\ \text{M} \times \frac{10 \times 10^{-6}\ \text{l}}{3 \times 10^{-3}\ \text{l}} = 1.66 \times 10^{-8}\ \text{M}$$

However, the enzyme has four active sites per molecule so the active site concentration is given by:

$$1.66 \times 10^{-8}\ \text{M} \times 4 = 6.64 \times 10^{-8}\ \text{M}$$

Sample Calculation of the Turnover number (TON)

$$\text{TON} = \frac{\mu\ \text{moles product/min at saturation}}{\mu\ \text{moles enz. active sites}}$$

$$= \frac{V_{max}}{\text{active site concentration}}$$

$$= \frac{1 \times 10^{-3}\ \text{moles per liter/min}}{6.64 \times 10^{-8}\ \text{Molar}}$$

$$= 1.5 \times 10^{4}/\text{min per active site}$$

Note: In these examples the magnitude of the data used was not intended to reflect that measured in the laboratory. Also, in order to minimize the number of units changed we have chosen to do all calculations in terms of moles per liter. v_o and V_{max} are often quoted in terms of the relevant assay volume i.e., μmole of product per ml per minute. Notice that TON was defined as μmoles of product per minute formed per μmole of enzyme. Since "μmoles" appears as units in both the top and bottom of the expression, it cancels, leaving the final unit, min.^{-1}. Thus, the value of the TON is the same whether one considers molecules of product per molecule of enzyme or moles of products per mole of enzyme. However, care must be taken to be sure the units of the product and enzyme concentrations are the same. K_m is always quoted in molar terms independent of the units chosen for V_{max}.

Sample Calculation for Unknown Pyruvate

Suppose the values of K_m and TON from the above example have been determined and the following data are obtained for an unknown pyruvate solution.

Pyruvate dilution	*Initial ΔA/min*	*v_0, moles per liter/min*
undiluted	1.69	27.3×10^{-5}
1 part in 3	0.66	10.6×10^{-5}
1 part in 10	0.23	3.7×10^{-5}

The Michaelis-Menten equation can be rearranged as:

$$v_0 = \frac{V_{max}\ [S]}{K_m + [S]}$$

$$[S] = \frac{v_0 K_m}{V_{max} - v_0}.$$

If the same enzyme concentration is used here as was used in the first example, V_{max} is known. Otherwise V_{max} must be calculated from the TON and enzyme concentration.

Dilution	*Pyruvate in cuvette*	*Correction for dilution in reaction mixture*[a]	*Correction for dilution of stock solution*[b]
none	3.86×10^{-4} M	2.89×10^{-3} M	2.89×10^{-3} M
1:3	1.30×10^{-4}	9.78×10^{-4}	2.93×10^{-3}
1:10	3.84×10^{-5}	2.88×10^{-4}	2.88×10^{-3}

[a]0.4 ml pyruvate diluted to 3.0 ml. So pyruvate in cuvette × 7.5 = the correction for dilution in reaction mixture.
[b]Correction is x1, x3, x10 for none, 1:3, and 1:10, respectively.

Sample Calculation of Unknown LDH

Suppose an unknown solution of LDH gives the data in the accompanying tabulation.

Dilution	*Initial ΔA/min*	*v_0, moles per liter/min*
None	3.05	49.2×10^{-5}
1:3	1.04	16.7×10^{-5}
1:10	0.31	5.0×10^{-5}

If we rearrange the Michaelis-Menten equation after substituting $V_{max} = E_T \bullet \text{TON}$ (E_T = total enzyme concentration),

$$E_T = \frac{v_0\ (K_m + S)}{(\text{TON})\ (S)}.$$

If S is much larger than K_m (at least 10 times larger), then

$$E_T = v_0/\text{TON}.$$

If we use this equation and remember that the TON was calculated per active site, not per enzyme molecule, the unknown LDH is calculated as follows:

Dilution	*LDH sites in cuvette*	*LDH in cuvette*[a]	*Correction for dilution in reaction mixture*[b]	*Correction for dilution of stock solution*[c]
None	3.28×10^{-8} M	8.20×10^{-9} M	2.46×10^{-6} M	2.46×10^{-6} M
1:3	1.11×10^{-8}	2.78×10^{-9}	8.35×10^{-7}	2.50×10^{-6}
1:10	3.33×10^{-9}	8.33×10^{-10}	2.50×10^{-7}	2.50×10^{-6}

[a]LDH in cuvette = LDH sites in cuvette/4.
[b]Correction for dilution in reaction mixture = LDH in cuvette × 300, i.e., 10 μl diluted to 3.0 ml.
[c]Correction for dilution of stock solution = Sample corrected for dilution in *reaction* mixture × 1 (none), × 3(1:3), × 10(1:10).

Selected References

Segel, I. H. 1975. *Enzyme Kinetics: Behavior and Analysis of Rapid Equilibrium and Steady State Enzyme Systems*. New York, Wiley.

Stryer, L. 1981. *Biochemistry*. 2nd ed. pp. 103–134, 352–353. San Francisco, Freeman.

Wolf, P. W., Williams, D., and Von der Muehel, E. 1973. *Practical Clinical Enzymology and Biochemical Profiling*. New York, Wiley-Interscience.

Problems

1. Calculate the concentration of lactate dehydrogenase (LDH) and its active sites, using the following sequence.
   a. Determine the absorbance at 280 nm of stock LDH.
   b. Calculate the concentration of stock LDH.
   c. Calculate the concentration of LDH after dilution in the reaction mixture.
   d. Calculate the concentration of LDH active sites.

2. Calculate the kinetic constants of LDH.

	[Pyruvate]	*[Pyruvate] diluted in reaction mix*	*1/S*	*Initial ΔA/min*	*Moles/liter/min*	*1/v$_o$*
a.						
b.						
c.						
d.						
e.						

Make a plot of 1/v$_o$ against 1/S on graph paper, and use the plot to determine the following kinetic constants: $-1/K_m$ (M^{-1}), $1/V_{max}$ (M^{-1} min), Slope (min), TON (min^{-1} per active site), K_m (M), V_{max} (M/min).

3. Use the following protocol to calculate the concentration of your pyruvate unknown.

Dilution	*Initial ΔA/min*	*Moles per liter/min*	*[Pyruvate] in cuvette*	*Correction for dilution in rx mix*	*Correction for stock dilution*
Undiluted					
1:3					
1:10					

Calculate the average value of the pyruvate unknown in the stock solution.

4. Use the following protocol to calculate the concentration of your LDH unknown.

Dilution	*Initial ΔA/min*	*Moles per liter/min*	*[LDH] sites in cuvette*	*[LDH] in cuvette*	*Correction for dilution in rx mix*	*Correction for stock dilution*
Undiluted						
1:3						
1:10						

Calculate the average value of the LDH unknown in the stock solution.

5. Why do we consider only initial velocities?

6. Is it possible to have too much or too little pyruvate in the unknown sample for an accurate measurement? What factors limit the accuracy at either extreme?

7. Why was pyruvate held at 5 mM in the assay for LDH? What would be the effect of using a ten-fold lower concentration?

8. Draw on graph paper the initial reaction velocity as a function of substrate concentration for an enzyme (that is not allosterically regulated) with only one substrate. Label V_{max} and K_m with arbitrary units.

9. What is meant by saturation of the enzyme in terms of the dependence of v_o on [S]?

10. K_m has units of concentration. What is the definition of K_m as a substrate concentration?

11. Lactate dehydrogenase has two substrates, NAD^+(NADH) and lactate (pyruvate). Kinetics for such a system are complex. How was this simplified for your experiment?

12. In your determination of the amount of LDH and pyruvate, the initial velocities were measured by continuously following the change in optical density. It is often possible to establish assay conditions (especially for enzymes that catalyze essentially irreversible reactions) so that the initial velocity is maintained for a relatively long time. When this is the case, the amount of enzyme present is directly proportional to the amount of product formed during a precisely timed interval. Thus a single determination of product at the end of the interval is sufficient to calculate the amount of enzyme present. When this type of assay is used in the clinical laboratory a reagent such as perchloric acid is added to the assay mixture at the end of the time interval. Explain why.

Tissue Distribution of Lactate Dehydrogenase Isozymes 3

John D. Lipscomb

Many enzymes are composed of multiple subunits that are identical in some cases and nonidentical in others. A few enzymes are active with either identical or nonidentical subunits as long as the final quaternary complex has the appropriate total number of subunits. This interchangeability of subunits leads to a family of enzymes within a single organism which catalyzes the same reaction. These isoenzymes, commonly called isozymes, have distinct differences in their physical properties (molecular weight, isoelectric point, denaturation temperature) as well as their catalytic properties (K_m, turnover number, pH optimum). However, since the subunits usually arise from genes produced by gene duplication from a single precursor, they share many structural similarities.

In this experiment we will study the isozymes of mammalian lactate dehydrogenase (LDH). We will use electrophoresis to separate the various LDH isozymes found in mammalian tissues. In addition, we will learn about the utility of serum LDH isozyme patterns as a diagnostic tool.

Principles

Isozymes are used by prokaryotes and eukaryotes to efficiently produce enzymes that can be fine-tuned to satisfy the metabolic requirements at a particular time in the life cycle or in a particular tissue. The catalytic properties of the enzyme can be altered simply by allowing more copies of one type of subunit to be produced rather than synthesizing an entirely different enzyme. In mammals, different isozymes frequently predominate in different tissues or are elaborated in various stages of development.

Function of LDH Isozymes

LDH is composed of four subunits which can be either type M, which is the major type in muscle tissue, or type H, which predominates in the heart and other highly aerobic tissue. Five isozymes are possible: H_4, H_3M, H_2M_2, HM_3, and M_4. All five are found in some tissues, such as kidney tissue. The fact that the subunits are interchangeable is easily demonstrated by dissociating the subunits in a mixture of purified H_4 and M_4. All five isozymes are elaborated in active form when the subunits are allowed to reassemble.

The biochemical rationale for the distribution of LDH isozymes in tissues follows from its catalytic properties. (See accompanying tabulation).

Isozyme	*TON*	*K_m(pyr)*	*Inhibition by pyruvate at physiological levels*
H_4	45,000 sec^{-1}	1×10^{-4} M	Yes
M_4	100,000 sec^{-1}	3×10^{-5} M	No

The H_4 isozyme is relatively sluggish and is inhibited by excess pyruvate (substrate). On the other hand, the M_4 isozyme has a much higher turnover number and is not easily inhibited. The metabolic function of LDH is summarized as:

1/2 glucose
↓ (NAD^+ → NADH; ADP → ATP)
pyruvate

lactate ← LDH — pyruvate ($NADH$ → NAD^+)

pyruvate → citric acid cycle ($4NAD^+$ → 4NADH); 4NADH → $4NAD^+$ with 2 O_2 → $4H_2O$, 12 ADP → 12 ATP

(+3 ATP from related reactions)

The cell can generate much more ATP (required for many cellular functions) by metabolizing pyruvate through the citric acid cycle. However, O_2 must be available or the NADH produced by the citric acid cycle cannot be coupled to ATP synthesis. In *aerobic tissues* such as the heart, O_2 is always in ample supply, so any pyruvate converted to lactate (away from the citric acid cycle) by LDH represents an energy loss. *Consequently the H_4 and H_3M isozymes with low catalytic efficiency predominate.* In tissue that must operate in a limited O_2 supply such as exercising muscle, the citric acid cycle cannot produce sufficient ATP. ATP is also produced during the conversion of glucose to pyruvate. This sequence of reactions is independent of O_2, and its overall rate increases manyfold when the ATP concentration decreases. However, during this process the limited supply of the cofactor NAD^+ is rapidly converted to NADH; thus the reaction stops. To overcome this problem in such tissues, NAD^+ is regenerated during the conversion of pyruvate to lactate by LDH. *Therefore in anaerobic tissues the more efficient M_4 isozyme is found almost exclusively.* One side effect of LDH activity is the accumulation of lactate in these tissues when the need for ATP is high. Large quantities of ATP are required for muscle contraction, and the buildup of lactate is responsible for the pain that accompanies vigorous exercise.

Experimental Detection of LDH Isozymes

The detection of the LDH isozyme pattern in serum is potentially difficult because the protein content of serum is very high. Staining techniques have been developed, however, that utilize the specificity inherent in the enzymatic reaction. In the first part of the experiment, the LDH isozymes in tissue extracts will be separated by electrophoresis on cellulose acetate strips. The strips will then be incubated in a staining solution containing NAD^+, lactate, and two dyes. LDH, like other enzymes, catalyzes the reaction in either direction. When lactate and NAD^+ are present in great excess, pyruvate and NADH are produced.

$$\underset{\text{lactate}}{CH_3-\overset{\overset{OH}{|}}{\underset{\underset{H}{|}}{C}}-\overset{\overset{O}{/\!/}}{C}-O^-} + NAD^+ \xrightleftharpoons{LDH} \underset{\text{pyruvate}}{CH_3-\overset{\overset{O}{\|}}{C}-\overset{\overset{O}{/\!/}}{C}-O^-} + NADH + H^+$$

The NADH is oxidized to NAD^+ by one of the dyes, phenazine methosulfate (PMS), which in turn is oxidized by the second dye, nitro blue tetrazolium (NBT).

$$NADH + H^+ + PMS \rightarrow NAD^+ + PMS_{[reduced]}$$

$$PMS_{[reduced]} + NBT \rightarrow PMS + NBT_{[reduced]}$$

The tetrazolium dyes are useful because they are blue when reduced (most dyes bleach) and the reduced dye is not oxidized by O_2. Thus the locations of LDH isozymes on the cellulose acetate strip are marked by a permanent blue stain.

In the second part of the experiment, the formation of hybrid isozymes from purified M_4 and H_4 will be demonstrated. Any procedure that separates the subunits in the solution can be used, but care must be exercised not to irreversibly denature the subunits themselves. One gentle technique that works with LDH is slow freezing and thawing. Usually enzymes tolerate freezing, but LDH quarternary structure is unstable in the ice lattice. Activity returns after thawing, but the statistical isozyme pattern is observed instead of the pure H_4 and M_4 pattern.

Quantification of the LDH isozyme pattern can be done in several ways. The cellulose acetate strips can be scanned with a densitometer and the resulting peaks integrated. The assumption that the intensity of the stain is proportional to the enzyme present may be only approximately true. A second method, faster and more precise, is frequently used to estimate the amount of H_4 isozyme relative to the isozymes containing the M subunit. The H subunit is much more stable to denaturants such as urea or heat. M_4 is destroyed by a 20 minute incubation with 2.5 M urea or by heating at 60°C for 15 minutes. In the final part of the experiment, the H_4 content of normal serum and serum after myocardial infarction will be estimated by differential stability in urea.

Clinical Applications

Identification of isozymes and development of techniques to discern rapidly the isozyme distribution (or pattern) form the basis for several clinical procedures that can reveal the site or stage of development of tissue damage. One of the most widely used clinical procedures assesses the isozyme pattern of lactate dehydrogenase (LDH) in the serum.

The level of LDH activity in the serum is normally quite low. The isozyme pattern differs from individual to individual, but all five isozymes are usually present. The LDH level in actively metabolizing cells, on the other hand, is quite high since LDH plays an important role in the metabolism of carbohydrates. Injury or disease in a particular organ results in death and rupture of cells in that organ and consequent release of LDH, along with the other contents of the cell, into the bloodstream. The LDH activity in the serum can increase manyfold under these conditions; the LDH isozyme pattern can be used to identify the tissue where cell rupture has occurred. For example, myocardial infarction causes the serum to take on the isozyme pattern of the heart tissue, which is rich in H_4 isozyme, whereas infectious hepatitis results in a "liver pattern" high in M_4 isozyme. The serum isozyme pattern is used both as a diagnostic tool and as a means to follow recovery. *The rise in serum LDH level after tissue damage is rather slow, often taking hours, a fact which limits its use in rapid diagnosis.* Figure 3.1. shows the isozyme patterns found in normal serum and in serum from individuals suffering from various diseases.

Procedures

Electrophoretic Separation of Isozymes

1. Use tubes containing: tissue extracts from mammalian heart, liver, kidney, and leg muscle, pure H_4 and M_4, which are located near the electrophoresis apparatus in the laboratory. Each group will run an electrophoresis on every sample, using cellulose acetate support strips.

2. The electrophoresis apparatus is filled with tris-barbital buffer, pH 8.4. The cellulose acetate strips have been immersed in the same buffer. Place a strip on a saturation pad so that the notch in the strip is at the top and at the left. Blot with filter paper to remove excess buffer, and apply sample to the middle of the strip using a capillary tube. Only one application is required. The strips are wide enough to apply two samples, side by side, on each. As quickly as possible place the strip in the apparatus so that it does not dry out, and run the electrophoresis for 30 minutes at 200 V. Mark the positive end of the strip. Dip the strips in the LDH staining solution (21 mg NBT, 2 mg PMS, 100 mg NAD^+, 0.7 ml 60% Na lactate in 10 ml KPO_4 buffer pH 7.4). After stained bands appear, dip the strip in 5% acetic acid for 10 minutes to remove excess stain, and blot with a paper towel. It may be necessary to incubate the strips at 37°C to speed development of the stain.

3. Work in groups of four. Each person should do one strip with two samples. Be certain all the samples are represented within the group. Run the H_4 and M_4 together on one strip.

Demonstration of Hybridization

Mix 30 μl each of purified H_4 and M_4 in 1.5 ml of 0.9% NaCl, and add an equal volume of 1 M $NaPO_4$ buffer pH 7 containing 10 mM 2-mercaptoethanol. Freeze the sample slowly. The freezing procedure is critical for successful hybridization, and some experimentation may be needed to find the proper freezing conditions. Placing the tube near the front of a shelf in an upright freezer usually works well. After leaving the sample in the freezer for at least two hours, thaw it slowly in an ice bucket. After electrophoresis, described above, stain the strips and scan them with a densitometer. Estimate the amount of each isozyme present from the integrated intensity of each stained band.

Estimation of H_4 by Urea Stability

Samples of H_4, M_4, normal serum, and serum after a myocardial infarction are provided at the lab bench. Place 10 μl of each sample in cuvettes containing the following.

1. 2.8 ml 50 mM phosphate buffer, pH 7.4
   0.1 ml NADH, 4.8 mM
2. 2.8 ml of 2.79 M urea in phosphate buffer
   0.1 ml NADH, 4.8 mM

Cover the cuvettes and mix by inversion. Allow the cuvettes to incubate at room temperature. After five minutes the control samples in phosphate buffer can be assayed by using the following procedure. Place the cuvette in the spectrophotometer and read the initial absorbance at 340 nm. Then start the reaction by placing 0.1 ml of 20 mM pyruvate solution in the cuvette, mix quickly by inversion, and take absorbance readings at the shortest intervals possible until the initial rate is clearly established (one to three minutes). Plot the data and record the initial velocity. Allow the samples in urea buffer to incubate at room temperature for exactly 20 minutes. Repeat the assay on these samples.

Calculation

The exact calculation of H_4 and non-H_4 fractions would require measurement of activity at several time points since the H_4 does not survive entirely and M_4 does not denature completely. However, very little M_4 activity survives, so it is generally assumed to denature completely, simplifying the calculation. First calculate the fractional loss in activity of H_4 using the data from the purified H_4 sample. Assume that the activity remaining in the urea-denatured serum samples is due to H_4 only. Correct for the loss in H_4 activity using the fraction calculated above. Subtract the H_4 activity from the total activity in the undenatured sample to obtain the non-H_4 activity. Calculate the fraction of H_4 and non-H_4 activity in the sample.

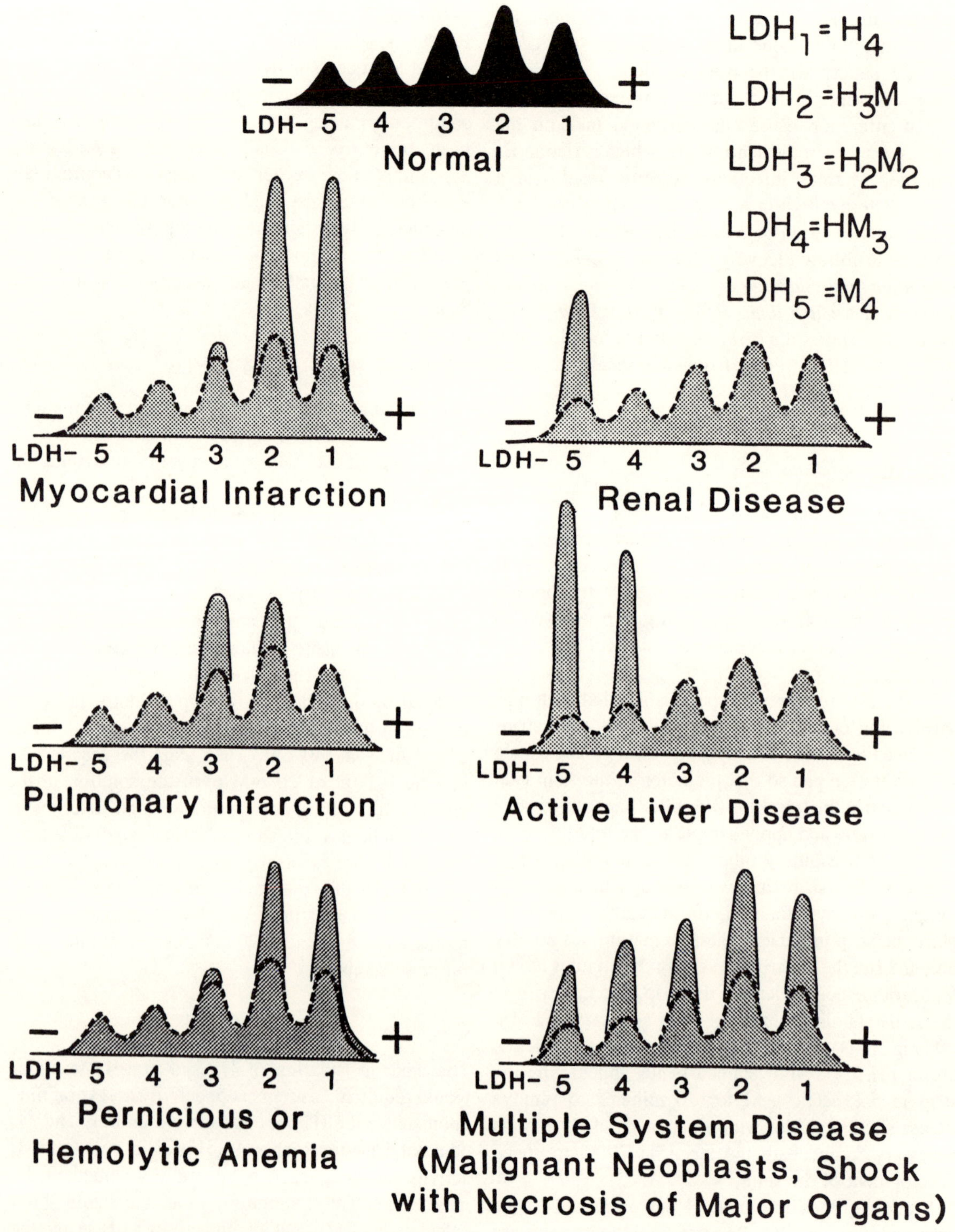

Figure 3.1. LDH isozyme patterns found in normal serum and in serum from individuals suffering from various diseases. Patterns from Gelman Technical Bulletin 23; LDH Isozyme Electrophoresis System.

A sample calculation follows:

Sample	Activity untreated (units/ml)[a]	Activity after urea treatment (units/ml)[a]	Fractional loss
H_4	10.0	7.0	0.30
M_4	10.0	0.0	1.00
Serum	17.0	8.0	0.53

[a]If only the fraction H_4 is desired, it is not necessary to convert initial velocity into units/ml.

Correct serum-sample residual activity for loss of H_4 due to urea treatment.

8/(1.0–0.30) = 11.4 units/ml
fraction H_4 in serum = 11.4/17.0 = 0.67
fraction non-H_4 in serum
= (17.0 − 11.4)/17.0 = 0.33.

Selected References

Cohn, R. D., Kaplan, N. O., Levine, L., and Zwilling, E. 1962. Nature and Development of Lactic Dehydrogenases. *Science* 136:962–969.

Dietz, A. A., and Lubrano, T. 1972. LDH Isoenzymes. In *Standard Methods of Clinical Chemistry*, ed. G. R. Cooper, Vol. 7, pp. 49–61. New York, Academic Press.

Everse, J., and Kaplan, N. O. 1975. Mechanisms of Action and Biological Functions of Various Dehydrogenase Isozymes. In *Isozymes II: Physiological Functions*, ed. C. L. Markert, pp. 29–43. New York, Academic Press.

Kaplan, A., and Szabo, L. L. 1979. *Clinical Chemistry: Interpretation and Techniques*. pp. 189–213. Philadelphia, Lea and Febiger.

Meyers, R. C., and Van Remortel, H. 1968. The Use of a Reagent Gel to Locate LDH Isozymes Separated on Cellulose Acetate Membranes. *Clin. Chem.* 14:1131–1134.

Problems

1. Sketch the isozyme patterns obtained from the following samples: heart, liver, kidney, muscle, M_4, and H_4.

2. What isozyme pattern would you expect brain tissue to have?

3. What is the observed percentage of isozymes in the hybrid mixture?

	H_4	H_3M	H_2M_2	HM_3	M_4
Observed	____%	____%	____%	____%	____%
Expected	____%	____%	____%	____%	____%

What factors could lead to differences in the observed and expected percentages?

4. Calculate the fractional loss of enzyme activity after urea treatment of the following samples, using the protocol given below.

Sample	Initial velocity (untreated)	Initial velocity (after urea treatment)	Fractional loss
H_4			
M_4			
Heart attack serum			

Serum v_o corrected for H_4 loss = ______

H.A. Serum v_o corrected for H_4 loss = ______

Fraction H_4 in serum = ______

Fraction non-H_4 in serum = ______

Fraction H_4 in H.A. serum = ______

Fraction non-H_4 in H.A. serum = ______

5. What problems are there in using the LDH isozyme procedure as a tool for diagnosis?

6. In the specific stain used here LDH converts lactate to pyruvate; in the spectrophotometric assays the reverse is true. What determines which way the enzyme drives the reactions?

7. Some enzyme reactions are termed irreversible. What would make some enzyme-catalyzed reactions reversible and others irreversible?

8. Many agents are used to exploit the differential stability of the H and M subunits of LDH. The most common are heat, pH, and urea. State why each of these procedures can denature enzymes.

9. One of the problems with attempting to quantitate a specific staining procedure such as that used in this experiment is that the intensity of the color produced depends upon the turnover number of the enzyme. Explain why this complicates direct quantitative comparison of H_4 and M_4 isozymes in the hybridization experiment.

Determination of Glucose in Serum and Urine 4

Esther F. Freier and John F. Van Pilsum

Glucose is utilized in one or more ways (as a metabolic fuel, in the synthesis of other metabolic fuels, in the synthesis of cellular components, etc.) by *all* tissues in the body. The measurement of levels of glucose in the blood (and urine) is helpful in the diagnoses of a large variety of diseases and is the most widely performed clinical biochemical analysis. Many different methods for determining glucose levels are currently being used by hospital clinical laboratories. The *specificities* of these methods for *glucose* vary greatly, as do the principles of the various procedures. You, as future physicians, should be aware of these facts in order to interpret correctly the glucose values of the patient. The purpose of this experiment is to demonstrate the principles and specificities of glucose methods presently being used.

In 1976 the College of American Pathologists conducted a survey of the methods used for determining glucose levels by 6,650 hospital laboratories in the United States. Table 4.1 provides the results of this survey, along with comparable data from 1974 and 1978. Note the trends in methodology used, particularly the shift from reduction methods to specific enzymatic procedures. The following glucose methods are examined in this laboratory exercise: ortho-toluidine (direct, manual), ferricyanide (automated), hexokinase (manual), glucose oxidase (PAP-automated and O_2 manual).

Table 4.1. Glucose Methods Used in the U.S. in 1974, 1976, and 1978

Method and System	Number of Laboratories 1974	1976	1978
Copper reduction—automated	1,154	436	147
Copper reduction—manual	48	21	—
Ferricyanide reduction	483	166	66
Glucose oxidase—Beckman O_2 analyzer	191	457	547
Glucose oxidase—PAP/automated	241	405	544
Glucose oxidase—PAP/manual	44	90	140
Glucose oxidase—other indicator	241	472	890
Hexokinase—automated	303	1,004	1,750
Hexokinase—manual	132	227	316
Ortho-toluidine—automated	166	129	556
Ortho-toluidine—manual, direct	2,842	2,257	1,580
Miscellaneous	655	818	147
	6,500	6,650	6,683

Principles

Ortho-toluidine Method

The ortho-toluidine procedure is a color reaction specific for the aldohexoses: glucose, mannose, and galactose. Aldohexoses other than glucose are usually, but not always, present in blood in very small amounts; therefore results obtained by this method approach the true value for glucose. The aldohexoses react with ortho-toluidine ($CH_3C_6H_4NH_2$) in glacial acetic acid to form a green chromogen that has a maximum absorbance at 625 nm.

In summary, the ortho-toluidine method is *not* specific for glucose, even though it is used by thousands of hospitals in the United States. You will demonstrate its lack of specificity by showing the extent of its reaction with glucose and galactose solutions of equal concentrations. In at least two inherited diseases, large amounts of galactose are present in the blood.

A protein-free filtrate of serum or plasma must be made before quantifying most of the nonprotein, water-soluble constituents in blood. This step is generally unnecessary for the determination of glucose levels by the ortho-toluidine procedure. Most hospitals use this procedure directly with plasma or serum. However, if the serum or plasma is obtained from blood in which hemolysis has occurred or if the serum contains large amounts of lipid or bile pigments, the glucose determinations by the direct ortho-toluidine procedure are falsely high. Under these conditions the removal of protein along with the lipid and bile pigments in the serum by protein precipitation reagents such as tungstic acid, perchloric acid, trichloracetic acid, etc., is recommended. The ortho-toluidine method will be performed manually as a direct procedure.

Ferricyanide Method

Glucose and other aldohexoses, ketohexoses, aldopentoses, and ketopentoses are called reducing sugars. That is, the aldehyde or ketone groups are easily oxidized by a variety of oxidizing agents, such as cupric sulfate or potassium ferricyanide. The extent to which these various oxidizing reagents are reduced by the blood has long been used as a quantitative method for determining blood-glucose levels. A great number of other substances in blood (e.g., vitamin C, uric acid, ketone bodies, creatinine) also reduce the oxidizing agents used in the glucose-oxidation procedures. The levels of these nonmonosaccharide-reducing substances in blood vary greatly in different diseases.

Used extensively in the past because it was easily automated, the ferricyanide method is based on the reducing property of glucose. Glucose reduces ferricyanide (yellow) in the presence of heat and alkali to ferrocyanide (colorless).

$$\underset{\text{ferricyanide}}{K_3Fe(CN)_6} \rightarrow \underset{\text{ferrocyanide}}{K_4Fe(CN)_6}$$

The decrease in the yellow color (read at 420 nm) is proportional to the amount of glucose but is also sensitive to other reducing substances such as creatinine, ascorbic acid, and uric acid. In the automated ferricyanide procedure, glucose is separated from protein by dialysis. The dialysis step also minimizes the effect of nonglucose-reducing substances, which migrate across the membrane at slower rates than glucose. The automated ferricyanide method will be demonstrated, and the extent to which it detects other reducing sugars and reducing compounds will be shown.

Hexokinase Method

Glucose is converted to glucose−6-phosphate in the presence of ATP and the enzyme hexokinase. The glucose−6-phosphate may then be oxidized to 6-phosphogluconic acid by the enzyme glucose−6-phosphate dehydrogenase with NAD as the cofactor. The amount of NADH formed may be quantified by the absorbance at 340 nm. The serum is incubated with ATP, hexokinase, NAD, and glucose−6-phosphate dehydrogenase, and the absorbance at 340 nm is determined after ten minutes. The specificity of the glucose−6-phosphate dehydrogenase minimizes reactivity with other hexoses. The reference method for glucose proposed by the U.S. Food and Drug Administration in 1974 is the hexokinase method performed on a protein-free filtrate of serum or plasma made with tungstic acid. This method will be performed manually as a direct procedure with blanks for specimen color. You will do the hexokinase procedure on serum and on equal molar solutions of galactose and glucose to illustrate the specificity of this procedure for glucose as compared with the ortho-toluidine method.

Glucose Oxidase (Colorimetric-Trinder PAP) Method

Glucose is oxidized to gluconic acid by glucose oxidase with the utilization of molecular oxygen and the production of hydrogen peroxide.

$$\text{Glucose} + O_2 + H_2O \rightarrow \text{gluconic acid} + H_2O_2$$

The hydrogen peroxide reacts with phenol and 4-aminophenazone in the presence of the enzyme peroxidase to form a red dye.

$$H_2O_2 + \text{phenol} + \text{4-aminophenazone} \xrightarrow{\text{peroxidase}} \text{red dye} + H_2O$$

The absorbance of the red dye is determined at 500 nm. The method, suitable for manual or automated procedures, will be demonstrated in the automated form. A number of naturally occuring substances in blood interfere in the above-mentioned oxidation of 4-aminophenazone to the red dye, and their interference will be demonstrated.

Glucose Oxidase (O_2 Consumption) Method

The oxygen utilized in the glucose oxidase reaction can be quantified by means of an oxygen electrode system such as the Beckman glucose analyzer. The rate of O_2 utilized is proportional to the glucose concentration in the serum. The H_2O_2 formed in the oxidation of glucose decomposes to $H_2O + O_2$.

$$2H_2O_2 \rightarrow 2H_2O + O_2$$

In this side reaction the production of oxygen decreases the net oxygen consumption. Thus the glucose values will be falsely low unless the H_2O_2 is removed. This is accomplished by adding catalase and ethanol to the reaction mixture.

$$H_2O_2 + \text{ethanol} \xrightarrow{\text{catalase}} \text{acetaldehyde} + H_2O$$

In addition, the complete removal of H_2O_2 is ensured by the presence of the molybdate and iodide in the reaction mixture.

$$H_2O_2 + 2I^- + 2H^+ \xrightarrow{\text{molybdate}} I_2 + 2H_2O$$

The oxygen-consumption method does not depend upon absorbance measurements; therefore direct measurements can be made on serum or plasma without interference from hemolysis, lipemia, or bile pigments (icterus). This method will be demonstrated.

The blood-glucose techniques discussed above have been evaluated against the FDA-proposed reference method with respect to reliability and specificity. The automated glucose oxidase, oxygen-rate method seems to be superior to all other procedures for both serum and urine. The reduction methods are unsatisfactory for urine, and the automated reduction methods relatively unsatisfactory for plasma.

Clinical Applications

The following are examples of disease states in which blood-glucose values are used as a diagnostic aid.

1. Diabetes mellitus
2. Endocrine disorders in which there is an excess secretion of: thyroxine (Graves' disease), growth hormone (acromegaly), and adrenal steroids (Cushing's disease)
3. Endocrine disorders in which there is a deficient secretion: of ACTH by the pituitary (hypopituitarism); of adrenal steroids by the adrenal gland (Addison's disease); of thyroxine by the thyroid gland (cretinism)
4. Pancreatic disease: acute pancreatitis (pancreatitis due to mumps) and chronic pancreatitis (cystic fibrosis)
5. Acute myocardial infarction
6. Chronic liver or kidney disease
7. Diseases of insulin receptor antibodies
8. Tumors: carcinoma of adrenal gland or stomach
9. Enzyme deficiency diseases: Von Gierke's syndrome, fructose intolerance, galactosemia

Physiological responses to challenges with glucose, insulin, and glucagon are evaluated in the clinic with the aid of glucose determinations for blood and urine. Some physiological responses to challenge with glucose are presented in Figure 4.1.

Three glucose tolerance curves are shown. The normal glycemic response to the oral administration of 100 g of glucose is illustrated by curve A. The rise in blood-sugar level is rapid, but a normal value is restored in two hours. The response observed in mild diabetes is shown by curve B. Normal blood-glucose values are not obtained until three hours after the ingestion of glucose. The fasting hyperglycemia and the high blood-glucose levels as long as three hours after the glucose ingestion are the results obtained in severe diabetes and are represented in curve C. Glycosuria usually occurs when the blood-sugar level is maintained (for several hours) above 160 mg per 100cc.

A variety of factors, other than diabetes, influence the glucose tolerance test:

1. *Carbohydrate starvation*

Starvation or the ingestion of high-fat diets de-

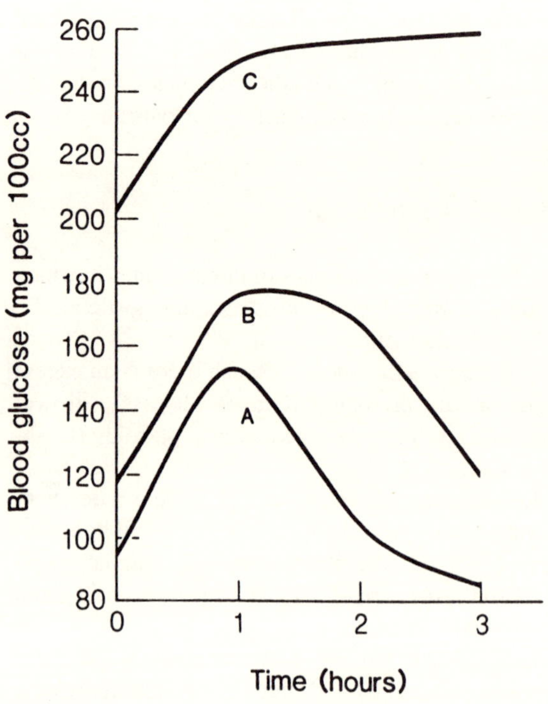

Figure 4.1. Some responses to challenges with glucose

presses the ability to utilize carbohydrate and results in hyperglycemia and glucosuria. Tolerance tests conducted under such conditions may elicit a typical diabetic response.

2. *Effect of exercise*

Strenuous exercise after the ingestion of glucose decreases the hyperglycemia and shortens the time required for the blood glucose to return to normal levels. Exercise just before the ingestion of glucose, on the other hand, exaggerates the hyperglycemia.

3. *Hyperinsulinism*

Benign or malignant tumors of the islets of Langerhans, functional hypertrophy and hyperplasia — all greatly increase the rate of utilization of carbohydrate and thus lower the glucose tolerance curves.

4. *Diseases of the liver*

Infectious hepatitis and other diseases of the liver inhibit the formation of glycogen and result in glucose tolerance curves resembling that of a diabetic.

5. *Acute and chronic infections*

Many infections depress the utilization of glucose, producing glucose tolerance curves of the type seen in diabetes.

6. *Nervous disorders*

A great variety of injuries to the brain depress the utilization of blood glucose and produce a diabeticlike glucose tolerance curve. Certain disturbances of the sympathetic nervous system increase the utilization of glucose and result in flat glucose tolerance curves.

Procedures

Analyze your serum and serum from a patient with galactosemia by manual glucose methods, i.e., hexokinase and ortho-toluidine. Be sure to include the known control-pool specimen to judge the quality of your analytical data. Work in pairs to perform the two manual methods. During stopping points, such as the heating step, visit the demonstrations of the automated ferricyanide reduction, the automated glucose-oxidase-PAP, and the Beckman O_2-glucose oxidase procedures. These automated methods are tested for specificity with other sugars, i.e., pure solutions of fructose, lactose, galactose, and xylose (all at concentrations of 200 mg/dl) and solutions of glucose (200 mg/dl) containing each of the following: creatinine (10 mg/dl), uric acid (10 mg/dl), and ascorbic acid (25 mg/dl). A normal urine and a normal urine to which glucose (100 mg/dl) has been added are analyzed for glucose by the automated methods. Students may analyze any of the above solutions for glucose by the manual methods if they wish. All reagents have been prepared for use.

Note that mg/dl is now the accepted terminology for mg% or mg/100 ml. However, the changeover to molar units of glucose is occurring in System International (S.I.) units.

100 mg glucose/deciliter = 100 mg% or
= 100 mg/100 ml or
= 5.5 m mol/liter

Ortho-toluidine

Procedure

The ortho-toluidine procedure is done on each of the following: the serum from each student, serum from a patient with galactosemia, four standard glucose solutions (50, 100, 200, and 300 mg/dl), the control serum (serum of known glucose concentration), and water (the colorimetric blank for the procedure).

1. Pipette *0.10 ml* of each sample (or solution) listed above into a separate screw-topped tube (a total of *9 screw-topped tubes will be needed*).

2. Add *6.0 ml* of ortho-toluidine reagent to each tube and mix well. Stopper the tubes with the Teflon-lined screw caps and heat the tubes in *a bath of boiling water for 12 minutes*. Remove the tubes from the water bath and place in an *ice bath for 5 minutes*. Remove from the ice bath and place in *a beaker of water at approximately 25 °C for about 10 minutes*.

3. Mix and transfer the colored solutions to *cuvettes* and record the *absorbance at 630 nm.* The solution used to set the colorimeter at 0 absorbance is the mix-

ture of water and ortho-toluidine. The color is stable for one hour.

Calculations

Plot absorbance versus glucose concentration of the standards. The standard curve should be linear and pass through the origin. Calculate the amount of glucose in the samples with the aid of the standard curve for glucose.

Hexokinase

This is a reagent kit manufactured by Smith Kline Instruments, Inc.

Components	*Concentrations*
Adenosine triphosphate	800 μM
NAD	830 μM
Mg^{++}	3 μM
Hexokinase	700 IU/l
Glucose-6 phosphate dehydrogenase	1,100 IU/l
Buffer (pH 7.5)	50 μM

Procedure

The hexokinase procedure is done on each of the following: the serum from each student, serum from a patient with galactosemia, one standard glucose solution (100 mg/dl), the control serum, and 0.9% saline (the colorimetric blank for the first portion of the procedure).

I.

1. Pipette 3.0 ml of the reaction mixture into each of the test tubes (10 mm × 100 mm).

2. Pipette 0.020 ml of each sample (or solution) listed above into one of the test tubes containing the reaction mixture. Transfer the reagent solution/saline mixture to an ultraviolet spectrophotometer cuvette and adjust the UV spectrophotometer to 0 absorbance at 340 nm. Allow the remaining reagent/sample mixtures to stand at least ten minutes at room temperature before recording the absorbance at 340 nm.

II.

Serum contains compounds that absorb light at 340 nm. Therefore the intensity of this absorbance must be determined and these values subtracted from the values obtained in part I of the procedure.

Add 3.0 ml of 0.9% saline solution to each of four test tubes (10 × 100 mm). Pipette 0.020 ml of the following into the test tubes containing the saline: the serum from each student, the serum from the patient with galactosemia, and the control serum. Mix and record the absorbance at 340 nm with the spectrophotometer set at 0 absorbance with 0.9% saline. The absorbance of these four solutions are the serum blanks aand must be subtracted from the absorbance of the solutions obtained in part I of the procedure.

Calculations

Calculate the mg glucose/dl in the serum by the following formulas:

$$F = \frac{\text{mg/dl glucose in standard solution}}{A_{340} \text{ of standard}}$$

unknown glucose $= F \times [(A_{340}$ of test sample) − (A_{340} of serum blank)].

Selected References

Cooper, G. R., and McDaniel, V. 1970. The Determination of Glucose by the Ortho-Toluidine Method. In *Standard Methods of Clinical Chemistry,* ed. R.P. MacDonald, Vol. 6, pp. 159–170. New York, Academic Press.

Dubowski, K.M. 1962. An o-Toluidine Method for Body Fluid Glucose Determination. *Clin. Chem.* 8:215–235.

Problems

1. Record mg glucose/dl (mg/100ml) for the three samples listed below, using the glucose procedures of ortho-toluidine and hexokinase.

Sample	*Ortho-toluidine*	*Hexokinase*
Student serum		
Control serum		
Serum containing galactose and glucose		

2. Current medical practice is to estimate insulin dosage on the basis of glucose excretion in the urine over a 24-hour period. Which of the procedures considered in this exercise are suitable for this purpose?

3. Which of the methods considered in this experiment do you consider the least specific for glucose? Which methods do you consider the most specific for glucose?

4. Explain why the two enzymatic and highly specific methods for glucose used in this laboratory experiment *might* be unsatisfactory for the blood or urine from certain patients. How would you determine if the values obtained by these methods on the urine or serum were or were not satisfactory?

5. A certain patient's serum might be expected to have unusually high levels of glucose. Using the enzymatic methods of glucose analysis, describe the simplest way for the clinical laboratory to ascertain that the reported value was correct.

Appendix

Demonstration of Glucose Determinations on Various Solutions by Three Types of Automatic Assays

Sample	Ferricyanide	Glucose Oxidase (PAP)	Glucose Oxidase (O_2 Consumption)
		mg/dl (mg/100ml)	
Normal urine	176	<5	<5
Normal urine + glucose (100 mg/dl)	275	96	100
Fructose (200 mg/dl)	206	<5	<5
Lactose (200 mg/dl)	100	<5	<5
Galactose (200 mg/dl)	164	<5	<5
Xylose (20 mg/dl)	228	<5	<5
Glucose (200 mg/dl) + creatine (10 mg/dl)	111	100	100
Glucose (200 mg/dl) + uric acid (10 mg/dl)	104	99	100
Glucose (200 mg/dl) + ascorbic acid (25 mg/dl)	114	26	100
Ascorbic acid (50 mg/dl)	31	<5	<5

Enzymatic Analysis of Blood Lipids 5

Ivan D. Frantz and John F. Van Pilsum

The lipids found in human blood are insoluble in water and must be solubilized and transported as complexes with proteins. The free fatty acids, which are added to the blood from adipose tissue, are transported as complexes with the serum albumin. The phospholipids, cholesterol (both free and esterified), and triglycerides are transported as protein-containing complexes called lipoproteins. The amounts of lipoproteins in the blood are a function of the nutritional state of the individual and have also been implicated in the etiology of certain disease states such as atherosclerosis. Alterations in the amounts of the various lipoproteins in blood are diagnostic indicators of several diseases or nutritional states. High levels of lipoproteins in the blood are called hyperlipoproteinemias. Some of these disorders are familial, and others are the result of diet or other environmental factors. Determination of levels of the various blood lipids is now routine practice in preventive medicine. Blood lipid levels are monitored while attempts are made to maintain their levels within normal ranges by a variety of techniques such as diet, exercise, drug therapy, and surgery.

Principles

Classification and Properties of the Blood Lipoproteins

The lipoproteins have been classified on the basis of their density: *high-density lipoproteins* (HDL) (d = 1.063 – 1.210); *low-density lipoproteins* (LDL) (d = 1.019 – 1.063); *very low-density lipoproteins* (VLDL) (d = 0.95 – 1.006); and *chylomicrons* (d < 0.95). The densities of the lipoproteins are determined by observing the rate of their sedimentation from plasma (adjusted to varying densities with KBr) while the solutions are exposed to a force of 105,000 × gravity in an ultracentrifuge. The reason the lipoproteins have different densities is that their chemical compositions vary greatly. The densities of the lipoproteins are directly related to the protein-to-lipid ratio. Some properties and compositions of the four classes of lipoproteins are illustrated in Figure 5.1. The lipoproteins also differ from each other in the size and nature of their apoproteins. The five distinct types of apoproteins that have been isolated from the different classes of lipoproteins are called apo A, B, C, D, and E. A high degree of specificity has been found for the presence of the individual apoproteins in the various classes of lipoproteins.

The four classes of lipoproteins migrate at different rates in an electric field. The HDL, LDL, and VLDL lipoproteins migrate with the α-, β-, and pre-β-globulin fractions, respectively (Figure 5.1), whereas the chylomicrons do not migrate under these conditions.

Biosynthesis and Functions of the Blood Lipoproteins

Chylomicrons

Chylomicrons are particles of newly absorbed (dietary) lipid, normally found in plasma one to four hours after a meal. They are formed from dietary triglycerides and cholesterol in the intestine and enter the bloodstream via the lymph system and the thoracic duct. Most of the chylomicron triglycerides are hydrolyzed to fatty acids and glycerol at the capillary walls of adipose tissue by the enzyme lipoprotein lipase (LPL). The fatty acids enter the adipocytes, where they are esterified to triglycerides and stored for future use. The remainder of the chylomicron particles (after the action of lipoprotein lipase) are rich in cholesterol and are called chylomicron remnants.

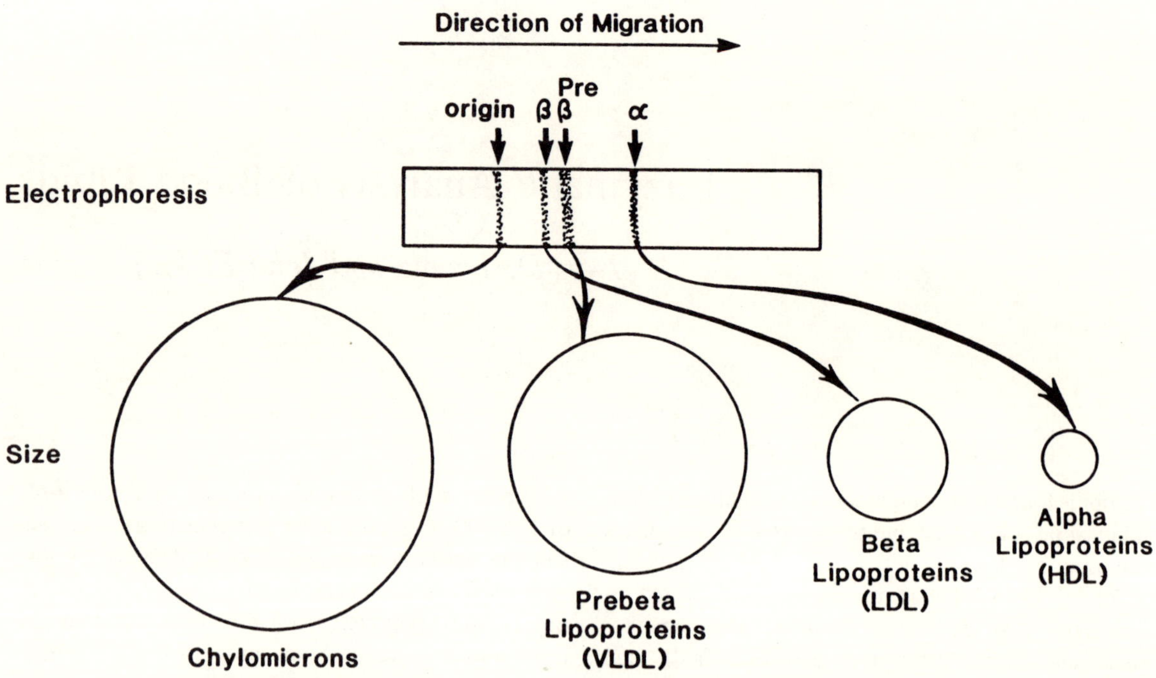

Figure 5.1. Some properties of the blood lipoproteins

Densities by ultracentrifugation	0.95 g/ml	0.95 – 1.006 g/ml	1.006 – 1.063g/ml	1.063 – 1.21 g/ml
Electrophoresis	Origin	Prebeta region	Beta region	Alpha region
Composition				
Protein	0.5 - 2.5%	10%	21%	50%
Phospholipid	3 - 15	22	22	26
Cholesterol	2 - 12	13	45	18
Free	(Less than	8	8	3
Esterified	half esterified)	5	37	15
Triglyceride	79 - 95	55	11	6
Size	0.1 - 5.0μ	M.W. 5 X 10^6–10^7	M.W. 1.3–3.2 X 10^6	M.W. 1.6–4.0 X 10^5

VLDL

These lipoproteins transport triglycerides that are synthesized in the liver and secreted into the blood. Practically all the triglycerides in plasma that are not in the chylomicrons are in the VLDL. VLDL is converted to LDL in blood when the triglycerides are removed by the action of lipoprotein lipase.

LDL

LDL is a principal carrier of the cholesterol deposited in the walls of the arteries. LDL is removed from the plasma by liver and extrahepatic tissue (e.g., adrenals and adipose tissue). LDL particles are taken up by specific cell-membrane receptors and then degraded to amino acids, cholesterol, and fatty acids in the lysosomes. Cholesterol diffuses from the lysosomes, where it is converted to its ester by the enzyme fatty acyl CoA:cholesterol acyltransferase (ACAT). Cholesterol ester is then stored in the cells.

HDL

HDL is synthesized in the liver and intestine. It removes cholesterol from peripheral tissues and returns it to the liver, where it is metabolized and excreted. HDL accumulates cholesterol in the form of cholesterol esters. Cholesterol esters in HDL are synthesized from cholesterol and phosphatidylcholine by the action of lecithin:cholesterol acyltransferase (LCAT).

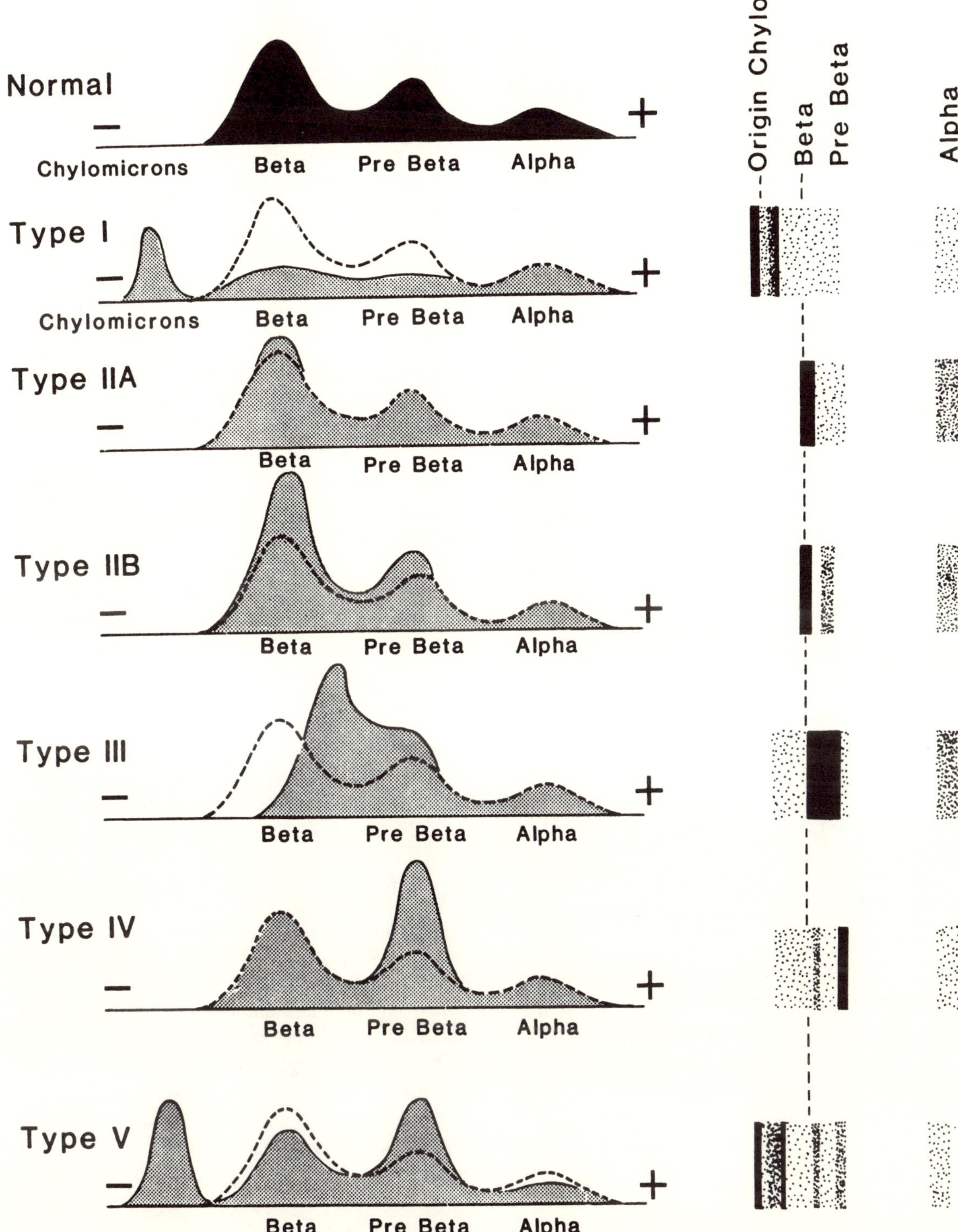

Figure 5.2. Relative amounts of blood lipoproteins in normal blood and in blood from patients with hyperlipoproteinemias.

Clinical Applications

Description of the Hyperlipoproteinemias

A schematic representation of the relative amounts of blood lipoproteins in normal blood and in blood from patients with hyperlipoproteinemias is shown in Figure 5.2.

Type I

This is a rare familial disease of an autosomal recessive type, in which the enzyme lipoprotein lipase is deficient in the adipose tissue. Individuals with this condition therefore cannot hydrolyze the triglycerides of the chylomicrons to fatty acids for deposition in the adipose tissue. High levels of chylomicrons present in the blood can be detected by electrophoresis and the standing plasma test. The absence of the enzyme lipoprotein lipase can be confirmed by the intravenous heparin injection test. Low amounts of the β-lipoproteins (LDL) and pre-β-lipoproteins (VLDL) are also observed after electrophoresis of plasma from Type I hyperlipoproteinemia patients.

Type IIA

This familial disease is transmitted by a single autosomal dominant gene. Abnormally high levels of β-lipoprotein (LDL) and its associated cholesterol are found in these individuals, a result of a deficiency of LDL receptors on cell membranes.

Type IIB

Abnormally high levels of both β-lipoproteins (LDL) and pre-β-lipoproteins (VLDL) are found in these individuals. Large amounts of LDL and VLDL lipoproteins are associated with high levels of both cholesterol and triglycerides in plasma. Patients with Type II (A or B) hyperlipoproteinemias have a high risk of developing atherosclerosis.

Type III

In this rare familial disease an abnormal lipoprotein called "floating β" is found in the blood (p.45). The "floating β" lipoprotein is also often found in the blood of hypothyroid individuals. Patients with type III hyperlipoproteinemia have a form of premature atherosclerosis with a high risk of developing peripheral vascular disease.

Type IV

Abnormally large amounts of pre-β-lipoproteins (VLDL) and associated triglycerides are observed in the blood of these patients. Type IV hyperlipoproteinemia is inherited as an autosomal dominant trait but also occurs in individuals with mild diabetes, obesity, and hyperinsulinemia. Type IV also is often found in alcoholics. Individuals with Type IV have a high incidence of coronary artery disease and peripheral vascular disease.

Type V

This rare disease is not well understood. Abnormally high levels of chylomicrons and VLDL particles (pre-β) are found. Some investigators think that the VLDL particles are so large that they behave like chylomicrons in electrophoresis. A Type V condition is sometimes found in patients with diabetes mellitus, chronic pancreatitis, liver, and kidney disease.

High levels of HDL

Large amounts of HDL in the blood have been associated with a *decreased* risk of developing atherosclerosis and a tendency to longevity. The suggestion has been made that the high levels of HDL prevent the accumulation of cholesterol in the arteries. The HDL level is slightly higher in most women than in men. It is often elevated in athletes and alcoholics.

Procedures

The total cholesterol and the cholesterol present in the HDL fraction of the plasma are determined. The VLDL and LDL are precipitated from the plasma by adding magnesium ions and heparin. The cholesterol in the HDL remaining in the supernatant is determined. The total triglycerides in the plasma also are measured. The triglyceride value is used to calculate the amount of cholesterol in the VLDL and LDL fractions. This calculation is based on the finding that VLDL cholesterol (in mg/dl) is approximately equal to 20% of the total triglycerides in the plasma (in mg/dl) (see Figure 5.1).

Cholesterol

Principle

The total cholesterol in the plasma and the cholesterol in the HDL fraction of the plasma are both determined

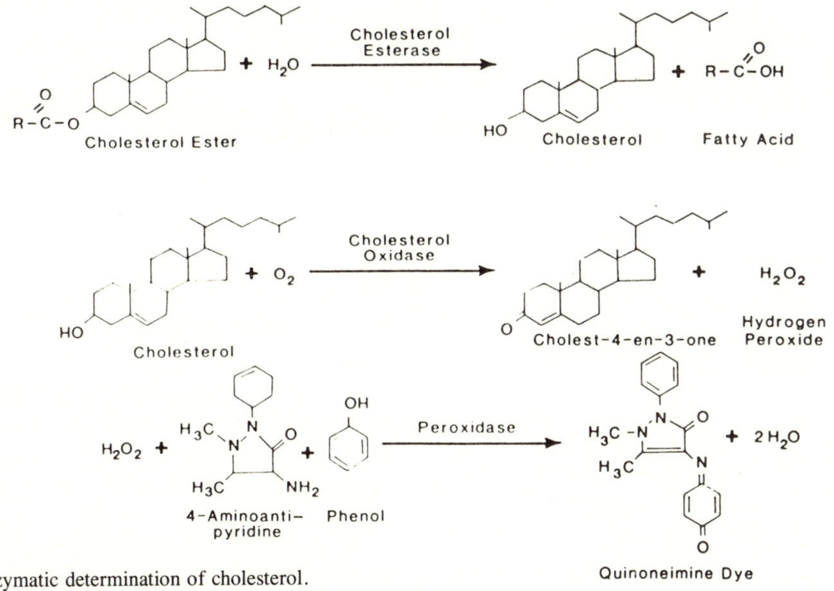

Figure 5.3. The enzymatic determination of cholesterol.

with a reagent kit that is based on three enzymatic reactions (see Figure 5.3). The colored product is measured at 500 nm, and its absorbance is directly proportional to the concentration of total cholesterol.

Reagents

1. Cholesterol reagent—from a reagent kit manufactured by Boehringer Mannheim Diagnostics, Inc.

Components	*Concentrations*
3-4-Dichlorophenol	0.78 mmoles
Phenol	1.09 mmoles
4-Aminoantipyrine	0.18 mmoles
Cholesterol oxidase	> 45 units
Cholesterol esterase	> 72 units
Peroxidase	> 36 units

2. Working precipitating solution (precipitates LDL and VLDL). Mix 10 ml manganese chloride, 1.06 M (20.98 g $MnCl_2 \bullet 4\ H_20$ in 100 ml distilled H_20), and 0.6 ml heparin (40,000 units per ml).

3. Standards. Preciset™ Cholesterol Standards, manufactured by Boehringer Mannheim Diagnostics, Inc., are used. They contain 50, 100, 200, and 300 mg cholesterol/dl in aqueous solution.

Procedure

Each group of four students determines the total cholesterol and the HDL cholesterol level of each plasma sample, and the total cholesterol of a control. The group establishes a standard curve, using 50, 100, 200, and 300 mg cholesterol/dl concentrations.

1. Pipette 1.0 ml of each plasma sample into a screwcap culture tube. Add 0.1 ml of working precipitating solution. Cap each tube and mix immediately on a vortex mixer.

2. Let tube stand for 10 minutes. Centrifuge at 2,500 rpm for 20 minutes. Pipette supernatants into test tubes, using a Pasteur pipette. The supernatants will be used to determine HDL cholesterol.

3. Place 10 μl of each sample (standards, plasmas, supernatants, and control) near the bottom of a micro cuvette. Place 10 μl of deionized H_20 into a cuvette for a reagent blank. Add 1.0 ml of cholesterol reagent to all cuvettes. Mix by inversion and let stand for 10 minutes at room temperature.

4. Read absorbance at 500 nm against the reagent blank. Plot absorbance against concentration of the standards. Calculate cholesterol concentration in the unknown samples by using the standard curve. For the HDL determinations, divide the mg cholesterol/d! (as determined by use of the graph) by 0.91 to correct for the dilution made in step 1.

Triglycerides

Principle

The reagents supplied in the kit contain buffer, ATP, magnesium ions, lipases, glycerol kinase (GK), glycerol−3-phosphate oxidase (GPO), peroxidase (POD), 4-chlorophenol, and 4-aminoantipyrine. The triglycerides in the sample are hydrolyzed into fatty acids and glycerol by lipase action.

$$\text{triglycerides} \xrightarrow{\text{lipase}} \text{fatty acids} + \text{glycerol}$$

The glycerol then reacts as follows:

$$\text{glycerol} + \text{ATP} \xrightarrow{\text{GK}} \text{glycerol-3-phosphate} + \text{ADP}$$

$$\text{glycerol-3-phosphate} + O_2 \xrightarrow{\text{GPO}} \text{dihydroxyacetone phosphate} + H_2O_2$$

$$2\ H_2O_2 + \text{4-aminoantipyrine} + \text{4-chlorophenol} \xrightarrow{\text{POD}} \text{quinoneimine} + \text{HCl} + 4\ H_2O$$

Concentration of triglycerides is proportional to the intensity of the color produced.

Reagents

A reagent kit manufactured by Stanbio™ Laboratory Inc. is used. The kit contains:

1. Enzymatic Triglyceride Standard GPO, 200 mg/dl.
2. Enzymatic Triglycerides Buffer GPO, PIPES buffer, 40 mmol/L, pH 7.5.
3. Enzymatic Triglycerides Reagent GPO (lyophilized). When reconstituted with buffer, the reagent contains the following:

Components	*Concentrations*
4-aminoantipyrine	0.4 mmol/l
4-chlorophenol	5.0 mmol/l
ATP	1.0 mmol/l
magnesium ions	5.0 mmol/l
lipases	150.0 U/ml
glycerol-kinase	0.4 U/ml
glycerol-3-phosphate oxidase	1.5 U/ml
peroxidase	0.5 U/ml
PIPES buffer solution (pH 7.5)	40 mmol/l

Procedure

Each group determines the triglyceride levels of the standard, of a control and of each of their plasma samples.

1. Place 10 μl of each sample (standard and plasmas) near the bottom of a semi-micro cuvette. Place 10 μl of deionized H_2O into a cuvette for a reagent blank.
2. Add 1.0 ml of triglyceride reagent to all cuvettes. Cover each cuvette with parafilm and mix gently by inversion. Let stand at room temperature for 10 minutes.
3. Read absorbance within 60 minutes at 500 nm, using the reagent blank to zero the spectrophotometer.
4. Calculate the mg triglycerides/dl of the plasma samples and control by the following formula:

$$\text{triglycerides (mg/dl)} = \text{absorbance of unknown} \times \frac{\text{concentration of standard (mg/dl)}}{\text{absorbance of standard}}$$

Calculations

Knowing the HDL cholesterol value and assuming that VLDL cholesterol is one-fifth of total triglycerides, you can determine LDL cholesterol by using the Friedwald equation:

LDL cholesterol = total cholesterol − (HDL cholesterol + VLDL cholesterol).

Use the data you obtained in the laboratory and sample worksheets 5.1 to 5.3 to calculate blood lipid values.

The normal distribution for plasma lipids and lipoproteins is shown in Table 5.1. In the event that any of your plasma lipid values are not within the normal range, further evaluation by your physician is suggested.

Sample Worksheet 5.1. Determination of Cholesterol

Cuvette	Specimen Name	Absorbance (500 nm)	Cholesterol (mg/dl)
1	H_2O (blank)		
2	Standard 50 mg/dl		
3	Standard 100 mg/dl		
4	Standard 200 mg/dl		
5	Standard 300 mg/dl		
	HDL Supernatants Enter Names		
6			
7			
8			
9			
	Student Plasma Enter Names		
10			
11			
12			
13			
14	Control		

Sample Worksheet 5.2. Determination of Triglycerides

Cuvette	Specimen Name	Absorbance (500 nm)	Triglycerides (mg/dl)
1	Standard 200 mg/dl		
2	Control		
3			
4			
5			
6			

Sample Worksheet 5.3. Calculation of Lipoprotein Fractions

	Measured by Student			Calculated	
Student Plasma	Total Triglycerides (mg/dl)	Total Cholesterol (mg/dl)	HDL Cholesterol (mg/dl)	LDL Cholesterol (mg/dl)	VLDL Cholesterol (mg/dl)
1					
2					
3					
4					
5 Control					

Table 5.1. Normal Distributions for Plasma Lipids and Lipoproteins[a]

			Cholesterol (mg/100ml)			
			HDL			
Age	Total Plasma Cholesterol[b] (mg/100 ml)	Plasma Triglycerides[c] (mg/100 ml)	Males	Females	LDL	VLDL
10–19	205	140	30–65	30–70	50–170	5–25
20–29	210	140	35–70	35–75	60–170	5–25
30–39	240	150	30–65	35–80	70–190	5–35
40–49	265	160	30–65	40–85	80–190	5–35
50–59	280	190	30–65	35–85	80–210	10–40

[a]Data compiled by the National Heart, Lung, and Blood Institute.
[b]Upper 10% cutoff
[c]Upper 5% cutoff

Selected References

Brown, M. S., Goldstein, J. L., and Fredrickson, D. S. 1983. Familial Type 3 Hyperlipoproteinemia (Dysbetalipoproteinemia). In *The Metabolic Basis of Inherited Disease*, eds. J. B. Stanbury, J. B. Wyngaarden, D. S. Fredrickson, J. L. Goldstein, and M. S. Brown, 5th ed., pp. 655–671. New York, McGraw-Hill.

Fredrickson, D. S., Levy, R. I., and Lees, R. S. 1967. Fat Transport in Lipoproteins-an Integrated Approach to Mechanisms and Disorders. *N. Engl. J. Med.* 276:34–44, 94–103, 148–156, 215–225, 273–281.

Friedewald, W. T., Levy, R. I., and Fredrickson, D. S. 1972. Estimation of the Concentration of Low-Density Lipoprotein Cholesterol in Plasma, Without Use of the Preparative Ultracentrifuge. *Clin. Chem.* 18:499–502.

Glomset, J. A., Norum, K. R., and Gjone, E. 1983. Familial Lecithin: Cholesterol Acyltransferase Deficiency. In *The Metabolic Basis of Inherited Disease*, eds. J. B. Stanbury, J. B. Wyngaarden, D. S. Fredrickson, J. L. Goldstein, and M. S. Brown, 5th ed., pp. 643–654. New York, McGraw-Hill.

Glueck, C. J. 1977. Classification and Diagnosis of Hyperlipoproteinemia. In *Hyperlipidemia: Diagnosis and Therapy*, eds. B. M. Rifkind, and R. I. Levy, pp. 17–39. New York, Grune and Stratton.

Goldstein, J. L. and Brown, M. S. 1983. Familial Hypercholesterolemia. In *The Metabolic Basis of Inherited Disease*, eds. J. B. Stanbury, J. B. Wyngaarden, D. S. Fredrickson, J. L. Goldstein, and M. S. Brown, 5th ed., pp. 672–712. New York, McGraw-Hill.

Hainline, A., Karon, J., and Lippel, K., eds. 1982. *Manual of Laboratory Operations: Lipids and Lipoprotein Analysis*. 2nd ed. Lipid Research Clinics Program, National Heart, Lung, and Blood Institute, National Institutes of Health and Human Services, Bethesda, Maryland.

Herbert, P. N., Assmann, G., Gotto, A. M. Jr., and Fredrickson, D. S. 1983. Familial Lipoprotein Deficiency: Abetalipoproteinemia, Hypobetalipoproteinemia, and Tangier Disease. In *The Metabolic Basis of Inherited Disease*, eds. J. B. Stanbury, J. B. Wyngaarden, D. S. Fredrickson, J. L. Goldstein, and M. S. Brown, 5th. ed., pp. 589–621. New York, McGraw-Hill.

Nikkilä, E. A. 1983. Familial Lipoprotein Lipase Deficiency and Related Disorders of Chylomicron Metabolism. In *The Metabolic Basis of Disease*, eds. J. B. Stanbury, J. B. Wyngaarden, D. S. Fredrickson, J. L. Goldstein, and M. S. Brown, 5th ed., pp. 622–642. New York, McGraw-Hill.

Searcy, R. L. 1969. *Diagnostic Biochemistry*. pp. 161–177, 356–372, 517–525. New York, McGraw-Hill.

Young, D. S., Pestamer, L. C., and Gibberman, V. 1975. Effect of Drugs on Clinical Laboratory Tests. *Clin. Chem.* 21:1D–432D.

Zannis, V. I., Just, P. W., and Breslow, J. L. 1981. Human Apolipoprotein E. Isoprotein Subclasses Are Genetically Determined. *Am. J. Hum. Genet.* 33:11–24.

Problems

1. A six-year-old child is found to have a plasma cholesterol concentration of 250 mg/dl and triglycerides of 3,000 mg/dl. When the plasma was allowed to stand overnight in the refrigerator, a creamy layer appeared at the top, with a clear infranatant. Discuss the diagnostic possibilities in this case and the additional information needed to decide among them.

2. A 30-year-old woman has a plasma cholesterol concentration of 750 mg/dl, with normal triglycerides. Receptors for low-density lipoproteins were found to be so low as to be barely detectable in her fibroblasts grown in tissue culture. What would you expect to find if you had the opportunity to measure blood lipids and LDL receptors in her parents and her five children?

3. Mention some of the additional examinations that would be indicated before the physician is justified in making a diagnosis of primary hyperlipoproteinemia in a patient with elevated blood lipids.

Appendix

Analytical Procedures for Plasma Lipids and Lipoproteins

Total Plasma Cholesterol

Historically, most methods used for the measurement of total plasma cholesterol have been based on chemical reactions that yield a colored product. More recently, enzymatic methods have gained popularity. At least 35 commercial kits are currently available for the measurement of cholesterol. The principles involved in two widely used methods are briefly described below.

1. *Analysis with the Liebermann-Burchard reagent*
Cholesterol, both free and esterified, is extracted from the serum with isopropanol. Aliquots of the extract are mixed with the Liebermann-Burchard reagent, which is a mixture of acetic anhydride, glacial acetic acid, and sulfuric acid. The colored product formed is measured in a colorimeter.
2. *Enzymatic analysis with cholesterol oxidase*
Cholesterol esters are hydrolyzed enzymatically. The free sterol is then oxidized to a ketone by the action of cholesterol oxidase. Addition of a peroxidase causes the hydrogen peroxide also formed in this reaction to react with phenol and 4-aminoantipyrine to produce a quinoneimine dye.

Total Plasma Triglycerides

Triglycerides are usually measured by determining the amount of glycerol released on hydrolysis. Both nonenzymatic and enzymatic methods are currently in use.
1. *Fluorometric methods*
Phospholipids, which also contain glycerol, are first removed by adsorption on zeolite. The triglycerides are saponified in alcoholic potassium hydroxide. The glycerol liberated is oxidized with periodate to formaldehyde. Addition of ammonia and diacetylacetone causes the formation of 3,5-diacetyl−1, 4-dihydrolutidine, which is fluorescent.
2. *Enzymatic methods*
In one method, four enzymatic steps are involved: hydrolysis to glycerol and free fatty acids; reaction of the glycerol with ATP to form glycerol-1-phosphate and ADP; a reaction of ADP with phosphoenolpyruvate to form ATP and pyruvate; a conversion of pyruvate to lactate with the utilization of NADH. A second method, used in this session, is described on pp. 41–42.

HDL Cholesterol

VLDL and LDL are selectively precipitated from the plasma by adding manganese or magnesium ions and heparin or dextran. Cholesterol in the HDL remaining in the supernatant is measured by one of the standard cholesterol methods. The cholesterol in the VLDL and LDL fractions can be calculated if the triglyceride content of the blood is determined. This is discussed on page 42.

Chylomicrons

The presence or absence of chylomicrons is determined either by allowing the plasma to stand overnight in the refrigerator and looking for a creamy layer on top, or by looking for a lipid-staining band that fails to move from the line of application on an electrophoretic strip.

Method of Analysis of Individual Lipoproteins

Ultracentrifugation

The density of plasma can be increased above the normal value of 0.95 by adding potassium bromide. If the plasma is then subjected to centrifugation, the lipoproteins that have a density less than the plasma will rise or float to the top of the centrifuge tube. If the density of the plasma is adjusted to 1.006 g/ml, the VLDL will rise to the top of the tube upon centrifugation. The plasma can then be adjusted to 1.063 g/ml, and the VDL will rise to the top of the tube upon centrifugation, leaving the HDL at the bottom of the tube. This type of ultracentrifugation can be done on either a preparative or an analytical scale.

Electrophoresis

HDL, LDL, VLDL, and chylomicrons can be separated by electrophoresis on a variety of solid supports including filter paper, agarose, cellulose acetate, and polyacrylamide gel (Figure 5.1).
1. *Paper and agarose electrophoresis*
Electrophoresis is done on whole plasma and on a fraction of plasma after ultracentrifugation at a density of 1.006. The fraction that rises to the top of the centrifuge tube is called 1.006 Top (T), and the fraction that migrates to the bottom of the centrifuge tube is called 1.006 Bottom (B). The electrophoretic patterns of whole plasma and the 1.006 T fractions are compared. If a protein fraction is observed in the 1.006 T fraction that has migrated at the same or slightly greater rate than the β-fraction of whole plasma, this fraction is called a "floating β" lipoprotein.
2. *Polyacrylamide gel electrophoresis*
In this procedure, conducted on whole plasma, the gel matrix serves as a molecular sieve and the migration of large molecules is slower than the migration of small molecules.
3. *Wieland and Seidel test*
This test is a combination of electrophoresis and a precipitation of VLDL done on whole plasma. After electrophoresis, the slide is immersed in a mixture of heparin and magnesium chloride that precipitates the

VLDL. The slide is stained to determine if the precipitated VLDL is in the β-region.

4. *Intravenous heparin test*

The test is used to determine the presence or absence of lipoprotein lipase in adipose tissue. Heparin, when injected into a human, stimulates the excretion and activation of the enzyme lipoprotein lipase from adipose tissue. This test is done if the plasma triglyceride concentrations exceed 400 mg/dl. The plasma samples obtained from a patient prior to and after an intravenous injection of heparin are subjected to electrophoresis and examined for the presence of chylomicrons.

5. *Sequence of the analytical procedures*

A sample of plasma from the patient who has fasted for 12 to 14 hours is analyzed for its content of total cholesterol and triglycerides. Above-normal levels of these compounds are interpreted to indicate the occurrence of one of a number of types of hyperlipoproteinemias. If hyperlipoproteinemia is indicated, the plasma concentrations of cholesterol in the HDL, LDL, and VLDL fractions are determined by precipitation of the VLDL and LDL from plasma with a solution of magnesium ions and heparin. In certain cases it is necessary to analyze the plasma by ultracentrifugation, electrophoresis, or the Wieland and Seidel test.

Table 5.2. The Fredrickson Classification of Hyperlipoproteinemia.

Diagnostic Indicators	Type I	Type II	Type III	Type IV	Type V
Appearance of plasma	Milky After overnight refrigeration cream layer on top, clear below	Clear	Clear or turbid	Clear or turbid After overnight refrigeration: turbidity remains	Milky After overnight refrigeration: cream layer on top, milky below
Prominent feature of lipoproteins electrophoretograms	Chylomicrons elevated	β-lipo proteins elevated	"Broad-β" band	Pre-β lipoproteins elevated	Chylomicrons and pre-β lipoproteins elevated
Triglyceride level	Elevated	Normal to elevated	Elevated	Elevated	Elevated
Cholesterol level	Normal to elevated	Elevated	Elevated	Normal to elevated	Elevated
PHLA (Post-Heparin Lipolytic Activity)	Generally low	Normal	Normal	Normal	Low or normal
Glucose tolerance	Normal	Normal	Often abnormal	Often abnormal	Often abnormal
Fat tolerance	Abnormal	Normal	Mildly abnormal	Usually normal	Abnormal
Risk accelerated arterial disease	Low	High	High	Some, but not so high as with Types II and III	Low

Source: Technical Bulletin No. 335-UV. Revised 1982. Table 1, p. 8. Sigma Chemical Company, St. Louis, Missouri. Based on information in Fredrickson, Levy, and Lees (1976). Reproduced with permission.

A System for Classifying Patients with Hyperlipoproteinemia

A system has been introduced for classifying patients with hyperlipoproteinemia into five rather distinct phenotypes. This delineation is based upon a number of factors, but foremost in the classification is the concentration of triglycerides and cholesterol as well as the electrophoretic behavior of serum lipoproteins. This system is called the Fredrickson classification of hyperlipoproteinemia (see Table 5.2).

Conditions Showing Abnormal Serum Triglyceride Levels

Abnormal serum triglyceride levels are likely to be encountered in a number of conditions, as indicated in the accompanying tabulation. Administration of certain drugs and medications has been shown to influence serum levels of triglycerides. A comprehensive review has been prepared by Young et al. (1975) and should be consulted for further information.

Conditions with increased levels of triglycerides	*Conditions with decreased levels of triglycerides*
Addison's disease	Abetalipoproteinemia
Coronary artery disease	Cachectic states
Cystic fibrosis	Hyperthyroidism
Diabetes mellitus	Kwashiorkor
Essential hyperlipemia	
Glycogen storage disease	
Hypothyroidism	
Liver disease	
Nephrotic syndrome	
Pancreatitis	

The Use of Recombinant DNA in the Detection of Genetic Abnormalities

6

Denise M. McGuire, Howard C. Towle, and Dennis M. Livingston

The structure and function of nucleic acids have in recent years assumed an increasingly important and exciting role in biomedical research. Through the techniques known as gene cloning, medically important products such as human insulin and growth hormone are now being produced in recombinant bacteria for treatment of patients. In addition, these advances are providing new ways to test DNA from individuals for genetic abnormalities and a better understanding of the molecular nature of such genetic disorders.

Principles

DNA, or deoxyribonucleic acid, is the molecular store of genetic information. This information is sufficient to encode the sequence of every protein that will be produced in every cell and at every stage of an organism's life. The DNA molecule also has a built-in mechanism to ensure that this genetic information is passed accurately from cell to cell during mitosis and from generation to generation.

In the past several years a new and exciting area of DNA research has begun, using what is generally known as gene cloning or recombinant DNA technology. Through recent technological advancements, it is now possible to isolate specific gene sequences, manipulate and modify these DNA molecules *in vitro*, recombine DNA molecules from more than one type of organism, and eventually reintroduce these modified genes into living cells. The potential for correcting certain inborn errors of metabolism caused by DNA mutations is now receiving serious attention. The hazards, limitations, and ethical implications of such research are still largely unexplored and the subjects of intense discussion.

One of the major tools of recombinant DNA research is the use of site-specific endonucleases (restriction nucleases) capable of splitting DNA into fragments by cleavage at highly defined sites. These restriction endonucleases are isolated from bacteria in which they help protect the organism from foreign DNA. Each bacteria produces its own specific endonuclease with a different base specificity. For instance, a common laboratory strain of *Escherichia coli* contains a nuclease that cleaves any DNA at the hexanucleotide sequence -GAATTC-. Incidentally,
-CTTAAG-
the bacteria protect their own DNA by chemically modifying the sequency recognized by their own restriction nuclease. By obtaining a battery of restriction nucleases with different specificities, cellular DNA can be fragmented in a highly defined fashion for gene mapping and isolation of specific fragments.

The ability to specifically fragment large DNA molecules is only the first step in recombinant DNA technology. The next step, and the heart of the research, involves the ability to isolate specific gene fragments. The power of the technique is incredible. In a human, estimates of the total number of genes are usually about 300,000. To isolate one unique gene from this large number by standard chemical means would be infeasible. Even if it were possible, to obtain significant quantities of a unique DNA fragment would necessitate starting with gram quantities of total DNA. Both of these problems can be circumvented using recombinant DNA technology.

The first step involves recombining DNA fragments obtained from restriction enzyme digestion of total DNA with an appropriate vector DNA. For bacterial hosts, plasmid DNA's—small extrachromosomal DNA molecules capable of self-replication and often carrying drug resistance—are commonly used. DNA fragments are joined to plasmid DNA at the restriction endonuclease termini. The recombinant molecules formed are introduced into bacteria in a process known as transformation. Bacterial cells that have taken up the recombinant plasmid DNA molecules are selected by their ability to grow in the presence of the appropriate drug. Once a mixed population of DNA fragments has been used to transform a culture of bacteria, individual bacteria can be cloned and grown separately. Since each bacterial cell will contain only one inserted DNA element, each clone will be specific for a given DNA fragment. Thus isolation of the complex mixture of DNA fragments is obtained. In addition, once a bacterial clone containing a specific DNA fragment has been identified, it can easily be grown in large quantities in a matter of days. Therefore essentially unlimited supplies of the cloned DNA fragment are available.

Clinical Applications

To date, one of the most practical applications of this new DNA technology has been the cloning of DNA fragments coding for several human polypeptide hormones. For example, at the present time insulin for treatment of insulin-dependent diabetes is obtained largely from porcine pancreases. Such insulin often leads to formation of antiporcine insulin antibodies in patients treated. Recently, however, the human gene coding for insulin has been cloned in a bacterial cell. The bacteria has been altered by genetic manipulation to produce large quantities of the human insulin. Testing of this insulin is currently under way, and in a matter of years large supplies of human insulin should be available for treatment of diabetes. A second example of a medically important protein being produced in recombinant bacteria is human growth hormone. This hormone is highly species-specific, so treatment of growth hormone-deficient children requires the human form of the hormone. Previously, this product was isolated from the pituitaries of human cadaver donors. Since only small quantities of growth hormone can be isolated from a single pituitary, a long-term shortage of this hormone for patient treatment exists. However, the gene containing the genetic information for growth hormone has recently been cloned into a bacteria. Such recombinant bacteria can produce virtually unlimited quantities of this important product. Clinical testing of the recombinant growth hormone is currently under way.

Restriction enzyme analysis of DNA is also being used for screening of certain genetic defects, which can be identified by unique restriction enzyme maps. Restriction enzyme recognition sites within and surrounding a given gene provide a new class of genetic markers that can be used to follow specific genetic alleles. For example, a polymorphic variation of a restriction enzyme site adjacent to the globin gene has been found to be tightly associated with the structural mutation hemoglobin S (sickle-cell hemoglobin). Such a polymorphism can be used to predict the occurrence of a normal or abnormal gene. This technology may be particularly useful in identifying heterozygotic carriers of certain genetic abnormalities that are otherwise silent. It holds great promise for the future in aiding our understanding of gene structure and expression.

Procedures

The first step in altering the genetic message of an organism is the isolation of the genetic material from the donor organism by the procedure described in today's exercise. To isolate DNA in a pure form, three things must be done: DNA must be released from the whole cells and cell membranes in a soluble form as a complex with proteins; the DNA-protein complex must be dissociated; DNA must be separated from the contaminating protein and RNA. At the same time a number of conditions leading to denaturation of the DNA (e.g., extremes of pH, high temperature, and low ionic strength) must be avoided. Two experimental procedures are described in this chapter for the isolation of DNA. Details for the isolation of DNA from *E. coli* and from your own blood are outlined.

Isolation of DNA from *E. coli*

In the Marmur procedure (*J. Mol. Biol.* 3, 208, 1961), the *Escherichia coli* cells are lysed and DNA is released by action of the anionic detergent sodium dodecyl sulfate (SDS). The DNA-protein complex is dissociated with a high concentration of sodium perchlorate, and to facilitate its removal, the protein is denatured with chloroform. The nucleic acids (both DNA and RNA) are then precipitated from solution with ethanol.

Obtain 25 ml of an *E. coli* cell suspension (containing 1 g cell paste in 0.15 M NaCl, 0.1 M EDTA, pH 8.0) and add 2.5 ml 25% SDS. Heat the solution for five minutes in a water bath maintained at 60°C. During heating, the solution will become increasingly viscous owing to cell lysis and the release of DNA. Cool in an ice water bath, add 6 m. 5 M $NaClO_4$ followed by 30 ml chloroform-isoamyl alcohol (24:1, v/v), and shake the mixture vigorously in an 8-oz. prescription bottle for five minutes (the mixture will turn white as a result of protein denaturation). Place the mixture in conical thick-walled glass centrifuge tubes and centrifuge for 10 minutes at maximum speed in a desktop centrifuge. Following centrifugation, you should have three layers; a lower alcohol layer, a solid interface of denatured protein, and an upper aqueous layer which contains the DNA. Carefully aspirate the upper layer, measure its volume and add two volumes of 95% ethanol.

Strings of DNA will appear and can be removed by winding them around a glass rod. Even though RNA will not "string out" in ethanol, it does precipitate and is caught on the DNA strands so, at this stage, you have both DNA and RNA. Digestion of the RNA by RNase and chromatography of the nucleic acid mixture on hydroxylapatite are frequently used methods of freeing the mixture of RNA while maintaining the structural integrity of DNA. In the first method, the nucleic acids are incubated with the enzyme under conditions that allow the digestion of RNA. The DNA is then precipitated with ethanol and recovered for subsequent experiments. In the second method, the nucleic acids are loaded on a hydroxylapatite column in low salt (0.2 M phosphate) and washed thoroughly. Single-stranded RNA will *not* stick under these conditions, whereas the double-stranded DNA remains bound until eluted with higher salt (0.4 M phosphate). Because of time constraints you will not actually carry out these purification procedures.

The ratio of DNA to RNA in a precipitate that we have prepared for you by *the procedure described above* will be determined in today's laboratory period. The reason you cannot carry out this procedure on your own preparation of DNA is that it must stand for several days to solubilize. Today's experiments will illustrate the differences in the stabilities of DNA and RNA in acid and alkali. DNA and RNA are both stable in ice-cold 0.5 N acid. RNA (not DNA) is hydrolyzed to nucleotides in alkali at 37°C, whereas both DNA and RNA are hydrolyzed to nucleotides with hot acid.

Pipette 1 ml of the nucleic acid solution (A_{260} = 10–15) into six thick-walled conical centrifuge tubes, add an equal volume of 0.6 N HCl, vortex and centrifuge for 15 minutes at top speed in a tabletop centrifuge. Discard the supernatant, which will contain any acid-soluble material. *Suspend* two pellets in 2 ml ice-cold 0.5 N HCl and leave on ice. *Suspend* two pellets in 2 ml 0.5 N NaOH and incubate at 37°C for a half-hour. *Suspend* two pellets in 2 ml 0.5 N HCl and incubate at 70°C for a half-hour. Chill all samples on ice, add 8 ml of ice-cold 0.5 N HCl, and centrifuge for five minutes at top speed in a tabletop centrifuge. Recover the supernatants, take them to the spectrophotometer, and the lab assistant or teaching assistant will read the A at 260 nm.

Sample Calculations

The A_{260} of hot acid hydrolysate = 2.50

A_{260} of 37°C alkaline hydrolysate = 2.0

A_{260} of cold acid hydrolysate = 0.08.

Total nucleic acid A_{260} = 2.50 − 0.08 = 2.42

RNA A_{260} = 2.00 − 0.08 = 1.92

The E° of DNA or RNA = 11,000.

Concentration of nucleotides =

$$\frac{2.42}{11{,}000} = 2.2 \times 10^{-4} \text{ molar total nucleic acid (nucleotide)}$$

$$\frac{1.92}{11{,}000} = 1.75 \times 10^{-4} \text{ molar RNA. (nucleotide)}$$

Moles of nucleotide =

$2.2 \times 10^{-4} \times 0.01\ l = 2.2 \times 10^{-6}$ moles.

Average molecular weight = 333.

Mass of nucleotides = $2.2 \times 10^{-6} \times 333 = 7.33 \times 10^{-4}$ g. Each ml of nucleic acid solution was derived from 0.033 g of cell paste.

$$\frac{7.33 \times 10^{-4} \text{ g nucleic acid}}{3.3 \times 10^{-2} \text{ g cell paste}} = 2.22 \times 10^{-2} \text{ g}$$

DNA + RNA per g cell paste, or, in other words, 2.22% of the cell paste is nucleic acids.

Using the data you have acquired, calculate the proportion of DNA and RNA in your samples, and the yield of DNA and RNA per gram of *E. coli*. (See Sample Worksheet 6.1.)

Sample Worksheet 6.1. Calculation of DNA and RNA Content of *E. coli*.

Nucleic Acid	Proportion in *E. coli* Sample	Yield per Gram of *E. coli*
DNA		
RNA		

Isolation of DNA from Whole Human Blood

Each group of four will have two tubes labeled #1 and two tubes labeled #2, thawed and ready for use. These tubes contain the lysed nuclei from your blood that were prepared earlier (at the time of blood drawing).

Precipitation of DNA from Nuclei

Twenty μl of 12.5% SDS and 100 μl of Proteinase K are added to the cone area of tube #1, and the mixture is heated in a water bath at 55°C for 60 minutes to hydrolyze the nuclear proteins. The tube is removed from the water bath, and 150 μl of 8.3 M ammonium acetate and 1 ml cold 95% ethanol are added and the contents mixed by gentle swirling. The strands of DNA should now form. It may be necessary to cool the mixture in an ethanol-dry ice bath for 15 minutes. The strands of DNA may be wound around a glass stirring rod and removed from the solution.

DNA Precipitation and Quantitation of DNA

Twenty μl of 12.5% SDS and 20 μl of 2.5 M NaOH are added to the cone area of tube #2, the contents are mixed, and the tube is placed in a 65°C water bath for 60 minutes to hydrolyze the nuclear RNA. The tube is removed from the water bath, and 250 μl of 1 M tris-HCl buffer, pH 8.0, are added with gentle mixing. The tube is centrifuged at 2200 RPM (1000 × g) for five minutes to precipitate unwanted particulate matter. Two hundred and fifty μl of the supernatant solution are transferred to a 12 × 75 mm test tube; then 250 μl of 5 M ammonium acetate and 1 ml cold 95% ethanol are added. The contents of this tube are mixed by inversion, and the tube is placed in a dry-ice ethanol bath for 15 minutes to precipitate the DNA. The DNA is collected by centrifuging the tube at 3000 RPM (1500 × g) for 15 minutes at 4°C. The supernatant solution is removed from the pellet of DNA with a Pasteur pipette. The pellet of DNA is dissolved by adding 500 μl of 0.2 M NaOH. Sometimes it becomes necessary to vortex this mixture and to heat it at 65°C for five minutes in order to solubilize the DNA. Two and one-half ml of double distilled water are added to the solution of DNA, and the absorbance of this solution at 260 nm will be determined by your teaching assistant.

Demonstration of the Detection of Inheritable Disorders through Analysis of Restriction Site Polymorphisms

Inheritable disorders are caused by mutations of genes (changes in the base sequence of a DNA molecule). Until recently, a genetic disorder could be recognized only by the symptoms (phenotype), rather than from a direct analysis of the gene in question (genotype). Direct examination of the DNA sequence of a suspected mutant gene copy would be difficult and laborious even by today's methodologies. In some specific cases, though, a fortuitous change has taken place so that the mutation is associated with a change in a sequence site recognized by a restriction endonuclease. The restriction-site difference (polymorphism) between "normal" and "mutant" gene is readily detected in DNA from lymphocytes and in DNA from fetal cells obtained by amniocentesis. In this demonstration we will analyze a polymorphism in the β-globin gene that is associated with sickle cell anemia.

The method by which such a polymorphism can be detected is called Southern blotting. DNA from a single individual is broken by a restriction enzyme that hydrolyzes DNA at a specific sequence that appears at numerous but specific places throughout the genome. The thousands of DNA fragments are electrophoretically separated by size in an agarose gel. The DNA is then transferred in place from the gel to a nitrocellulose or special commercial filter. To detect the one DNA fragment of interest among the multitude of fragments on the filter, the DNA is hybridized with radioactive cloned (pure) DNA that has previously been isolated by genetic engineering techniques. The radioactive DNA forms a double helix with its complement that is bound to the filter. The location of the bound DNA fragment is then visualized by exposure to X-ray film. A restriction-site polymorphism will change the size of the detected fragment and thus change its position in the gel.

The six steps in the procedure are as follows:

1. DNA from lymphocytes is prepared by a method similar to that described earlier in this chapter. Additional steps are used to remove contaminating proteins and RNA in order to ensure that the DNA is pure enough to be cleaved by the restriction enzyme.

2. Restriction enzymes recognize specific DNA sequences; most recognize four or six base sequences that are palindromic. The DNA is combined with the restriction enzyme in a reaction mixture buffer (containing Mg^{+2} ion) and incubated.

3. The restriction fragments are subjected to electrophoresis in an agarose (very pure agar) gel. The fragments separate by size; the smaller fragments migrate faster than the bigger fragments. After electrophoresis the DNA is visualized by staining with a dye (ethidium bromide), which fluoresces when combined with DNA. Because human DNA contains over three million base pairs, one restriction enzyme will cleave the DNA into many fragments. Consequently

the gel contains too many bands to be able to visualize any individual fragment.

4. The DNA fragments are denatured in the gel with alkali, and the gel is then blotted against a nitrocellulose or other commercial type of filter. The DNA diffuses from the gel and binds irreversibly to the filter in the order in which it was found in the gel.

5. The filter containing the denatured, restriction fragments is now probed (hybridized) with DNA complementary to the β-globin gene. In this case the probe DNA is the cloned β-globin gene obtained by genetic engineering techniques. The probe DNA is made radioactive by the addition of radioactive deoxynucleotides with DNA Polymerase I of *E. coli*. Thus the probe DNA is "replicated" *in vitro* to make a radioactive copy.

The radioactive probe DNA is denatured by heating and then hybridized (combined) with the filter containing the immobilized restriction fragments. The radioactive probe will find its nonradioactive complementary sequence and form a double helix on the filter. After hybridization the filter is washed to remove radioactive molecules that have not formed a double helix with the immobilized DNA.

6. The filter is placed next to a piece of X-ray film and exposed (usually for days or even weeks). The radioactive decay darkens the X-ray film at the position of the hybridized radioactive DNA.

Detection of Sickle-Cell Anemia through Analysis of Restriction-Site Polymorphisms*

Sickle-cell anemia results from a mutation that changes a glutamic acid residue (coded by the triplet GAG) for a valine residue (coded by GTG) at position 6 in the β-globin chain of hemoglobin. As a result, sickle hemoglobin tends to crystallize in red blood cells, the cells become less flexible and are removed by the spleen, and anemia results. The mutation of A to T in the base sequence of the β-globin gene eliminates a restriction site for the enzyme *Dde*I (as well as sites for other restriction enzymes). The sickle hemoglobin mutation can, therefore, be detected by digesting sickle-cell and normal DNA with *Dde*I and performing Southern blot hybridization. Normal DNA will generate two *Dde*I fragments of 201 and 175 base pairs, whereas sickle-cell DNA will generate only one fragment of 376 base pairs (Fig. 6.1).

Although the restriction enzyme *Dde*I can be used to detect the sickle-cell mutation, it is not ideal for routine hospital screening procedures because there are too many *Dde*I restriction sites in the β-globin gene and its immediately neighboring DNA. As a result, digestion of the DNA produces a rather large number of small DNA fragments that are relatively difficult to separate by gel electrophoresis. However, another convenient enzyme, *Mst*II, has been discovered. *Mst*II cuts at the sequence CCTNAGG, which occurs less frequently than the *Dde*I site CTNAG. *Mst*II generates from normal DNA a 1.1-kilobase β-globin gene fragment; in sickle-cell DNA this is replaced by a 1.3-kilobase fragment. Fragments of this size are more easily separated and recognized than the considerably shorter ones *Dde*I produces. For that very practical reason, *Mst*II has become the enzyme of choice for the direct detection of the sickle globin mutation.

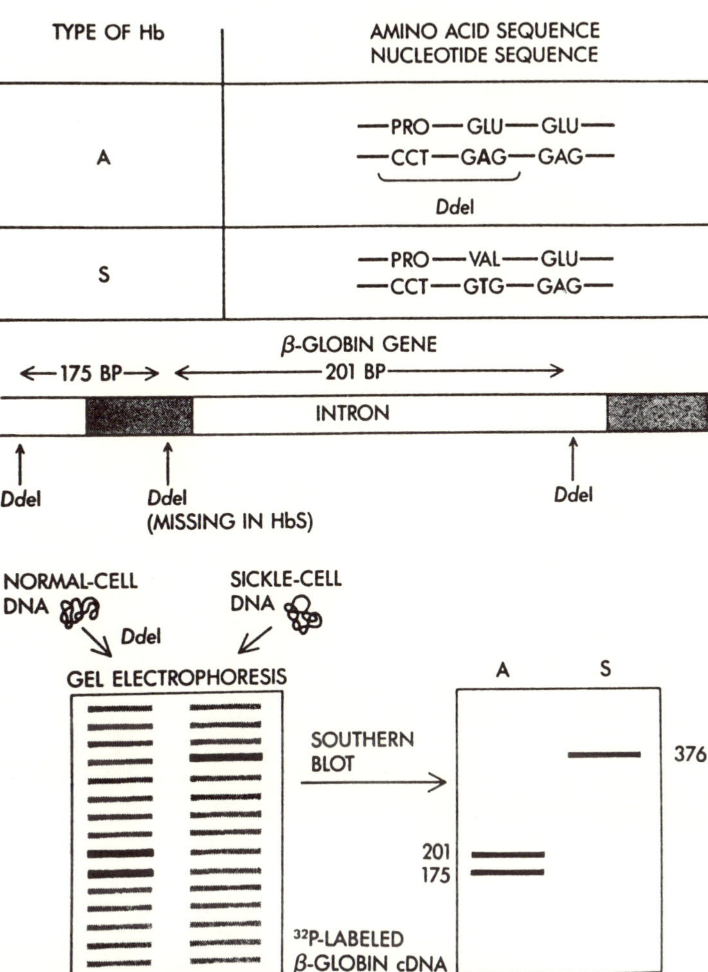

Figure 6.1. Detection of the sickle-cell globin gene by Southern blotting

*This section, including Figure 6.1, is reprinted from *Recombinant DNA: A Short Course* by James D. Watson, John Tooze, and David T. Kurtz, pp. 215–216. W. H. Freeman and Company. Copyright © 1983. Reproduced by permission.

Selected References

Kan, Y. W., and Dozy, A. M. 1978. Polymorphism of DNA Sequence Adjacent to Human β-Globin Structural Gene: Relationship to Sickle Mutation. *Proc. Natl. Acad. Sci. USA* 75:5631–5635.

Southern, E. M. 1975. Detection of Specific Sequences Among DNA Fragments Separated by Gel Electrophoresis. *J. Mol. Biol.* 98:503–517.

Problems

1. Calculate the DNA content of your nuclei and express it in terms of the number of base pairs per human nucleus, using the information below.

   a. 1 mg/ml DNA has an absorbance of 20 at 260 nm.

   b. 1 ml of blood contains 7.8×10^6 nucleated cells.

   c. The average molecular weight of a base pair is 666.

2. Give several explanations why your calculated value may differ from the known DNA content of human nuclei.

3. Explain why RNA is hydrolyzed by alkali, whereas DNA is not. Ribosomal RNA of eucaryotic cells contains about 3% of a modified nucleoside containing a 2′-O-methyl group on the ribose. What would be the alkaline digestion products of RNA containing this modification?

4. If a solution of DNA is gradually raised to a temperature of 85–90°C, an increase of about 30% is observed in the absorbance at 260 nm. What physical processes occurring as a result of this temperature change would be responsible for an increase in absorbance at 260 nm? Would RNA exhibit this phenomenon?

5. The restriction enzyme Hinc II cleaves at the recognition site (5 ′) GTPyPuAC, where Py can be either T or C, and Pu can be either A or G. What is the average distance between such cleavage sites expected in a random DNA sequence?

6. If a bacterial cell contains 1×10^{-14} grams of DNA as one molecule and if each base pair contributes 3.4 Angstroms to its length and has an average molecular weight of 660, what is the length of one molecule of DNA? Why should DNA of this length string out on a glass rod? Theoretically, how many proteins of molecular weight 40,000 can be coded for on this bacterial DNA? The average molecular weight of an amino acid is ~ 110.

7. If the size of the human genome is 3×10^9 base pairs (per haploid cell, e.g., sperm or ovum) and if a restriction enzyme that recognizes a four base palindromic sequence is used to cleave this DNA, what will be the theoretical average size of the fragments and how many will there be?

8. The DNA used as the probe in the polymorphism demonstration is a fragment of genomic DNA from a normal individual. Why does it hybridize to the DNA of other normal individuals as well as that of diseased individuals?

9. The DNA restriction enzyme digest pattern of a family with a history of sickle-cell anemia is shown in Figure 6.2. Sickle-cell anemia is a disease that is inherited as a recessive trait. The father F and the mother M are carriers of sickle-cell anemia.

   a. Why does the restriction enzyme pattern of the father and mother show three bands?

   b. From the DNA pattern of their children (C1, C2, C3), would you predict that any of the children would develop sickle-cell anemia? Would any of the children be carriers of the disease?

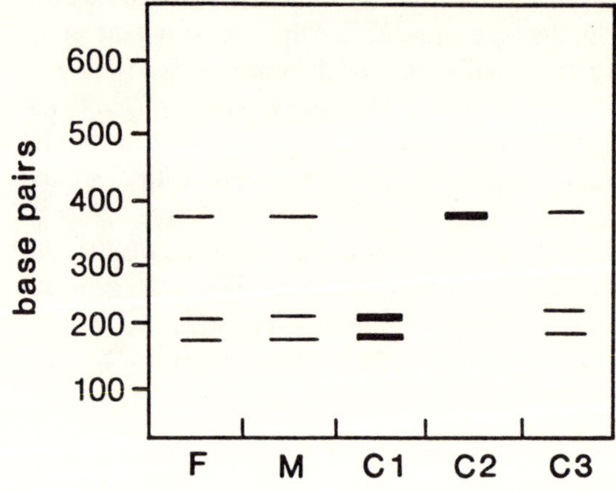

Figure 6.2. Southern Blot analysis of DNA

10. Sickle-cell anemia is a disease of the β-globin gene. In 87% of chromosomes with the sickle-cell allele, a change occurs in the sequence of DNA nearby such that a *HpaI* restriction site is absent. This loss of the *HpaI* site takes place in only 3% of normal chromosomes. (See Figure 6.3.). Speculate on why normal chromsomes are more likely to have the *HpaI* site than the sickle-cell chromosomes are. What genetic term is used to express the association of the restriction-site polymorphism and sickle-cell mutation? What genetic mechanism(s) would eventually make the probability of finding the restriction-site polymorphism equal for both normal and sickle-cell chromosomes?

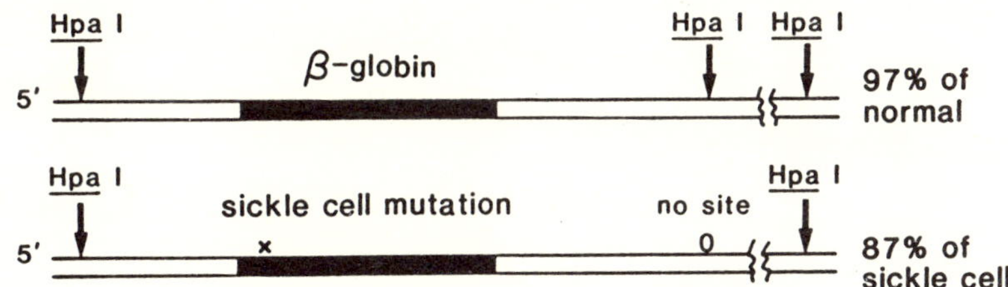

Figure 6.3. Location of Hpa I sites in the vicinity of the β-globin gene on normal and sickle-cell chromosomes

7 Inheritable Diseases and Genetic Engineering

Dennis M. Livingston

An expanding field of concern for the physician is the study of inheritable diseases and the genetic predisposition of individuals to contracting diseases. The purpose of this laboratory session is to describe: the genetics of inheritable diseases; how a genetic lesion results in a recognizable disease; and how health scientists can "treat" inheritable diseases.

Principles

Inheritance and Disease

All human cells except the gametes contain two of each of 23 *chromosomes*. One copy of each chromosome pair is inherited from the ovum and the other from the sperm. Which of the two copies of each pair is packaged into the gamete is a matter of chance. The chromosomes are, from the geneticist's point of view, a series of *genes* that code for proteins or important RNAs. If two genes from a chromosome pair are exactly alike, we are *homozygous* for that gene. On the other hand, if the two genes differ in any way, we then carry two different forms of the gene, called *alleles*, and are *heterozygous* for that gene. Different alleles occur because of *mutation*.

Inheritable diseases arise because some alleles of particular genes are defective in promoting the proper biochemical reaction. Fortunately, defective alleles can often be complemented by an effective allele of the same gene. If complementation occurs, the effective allele is said to be *dominant* and the defective allele is said to be *recessive*. Many genetic diseases result from receiving two defective alleles of the same gene, one from each gamete (parent). In this case the individual is homozygous for the recessive defective allele. Note that the parents can each be heterozygous, possessing one effective dominant and one defective recessive allele, and thus appear not to possess the defective trait. Of course the defective allele may be dominant and the effective allele recessive; the allele specifying Huntington's Chorea is one example of a dominant defective allele.

In humans 22 of the 23 pairs of chromosomes are called *autosomal chromosomes*. One pair of chromosomes, called *sex-chromosomes*, determines sex. Females have two X sex-chromosomes and thus possess two copies of the genes located on this chromosome. Males, on the other hand, have an X and a Y sex-chromosome. Either because the Y chromosome contains very few genes, or because most of the genes it does carry are never expressed, males functionally possess only one copy of the genes contained on the sex-chromosomes. This single functional copy must be inherited from the mother since the male's X sex-chromosome is received from the mother and the Y sex-chromosome from the father. The consequence of this in terms of an inheritable disease is that a *sex-linked* recessive trait carried by a heterozygous mother shows up in (half of) her sons, where only a single defective recessive allele is sufficient to specify the defective trait. Daughters of the heterozygous mother will not show the recessive sex-linked trait unless their father possesses that trait.

Biochemistry of Defective Alleles

Chromosomes are long molecules of DNA (with proteins added for structure and regulation). Thus the genes are specific regions or sequences of nucleotide bases along the DNA molecules. If the base sequences along the same region of the two DNA molecules on a chromosome pair are exactly alike, we possess the same allele of the gene on each chromosome and are homozygous for that gene. If, on the other hand, the nucleotide base sequence differs between the DNAs of the chromosome pair, we possess different alleles of the gene and are heterozygous for that gene. Differences in DNA sequence arise by mutation.

Each gene codes for a polypeptide (or important RNA such as rRNA and tRNA). The DNA sequence of the gene therefore contains the code for the amino acids in the polypeptide. Differences between DNA base sequence can result in differences in amino acid sequence of the polypeptide encoded by the gene. In addition the DNA sequence of a gene might also contain information for the regulation of the gene's expression.

Consider an enzyme comprised of a single polypeptide coded for by a single autosomal gene. Let us assume that for a functional enzyme we require a particular serine residue. If by mutation the DNA sequence of the gene is changed so that leucine is inserted in the protein instead of serine, the enzyme will be nonfunctional. For many individuals one leucine allele will not be deleterious because their second copy of the gene will more than likely contain a functional serine allele. The presence in the cell of both a defective and a functional enzyme usually results in an overall functional physiology. This is why most defective alleles are recessive. On the other hand, if an individual has two leucine alleles, he or she will suffer the consequence of the defect.

Clinical Applications

The social implications of prevention of inheritable diseases and of genetic engineering are so great that health scientists must understand the technicalities well and be able to explain the consequences and options available to the patients.

Examples of Inherited Diseases

Phenylketonuria

A typical autosomal recessive disease, phenylketonuria (PKU) occurs once in 26,000 births. Every child at conception obtains two copies of the gene coding for phenylalanine hydroxylase, one from the father, one from the mother. One can imagine that each drew two alleles from the "genetic grab-bag." The fact that one child in 26,000 randomly obtains two defective alleles demonstrates that the gene frequency for this defect is 1/160. (One chance in 160 for a defective allele on the first draw; one chance in 160 on the second draw; $1/160 \times 1/160 = 1/26{,}000$ for two draws.) The hidden heterozygous carriers in the population also drew twice, each time with a risk of 1/160. Therefore the carrier frequency is $1/160 + 1/160 = 1/80$. About three individuals in a medical class of 240 students are carriers of this defect. The probability that two carriers mate is $1/80 \times 1/80 = 1/6{,}400$. The probability that two carriers produce a defective child is one chance in four. This completes the cycle of computations since $1/6{,}400 \times 1/4 = 1/26{,}000$, which is the fraction of all children with PKU.

People with PKU are unable to degrade excess phenylalanine. For some unknown reason, the excess phenylalanine leads to mental retardation. PKU can be screened at birth by chemical analysis of the urine and blood. By limiting the amount of phenylalanine in the diet, the effect of the disease can be greatly reduced. Limitation of phenylalanine is usually unnecessary after the age of six, when the nervous system is better developed.

Tay-Sachs Disease

Tay-Sachs is another autosomal recessive disease. The defective enzyme is β-N-acetylgalactosaminidase; loss of the enzyme results in the accumulation of a ganglioside lipid within the nervous tissue. Death occurs at an early age.

The frequency of the defective allele in the United States is 1.3×10^{-3} so that the frequency of a homozygous recessive individual is 1.7×10^{-6}. This would mean an incidence of fewer than ten cases per year. However, the frequency of the defective allele is 12.6×10^{-3} among the American Jewish population and even higher for Jews of Eastern European descent. Individuals suspected of carrying the defective allele as heterozygotes can be screened and the heterozygotes identified because they have an intermediate level of the enzyme β-N-acetylgalactosaminidase. This knowledge is of value to the genetic counselor. In Tay-Sachs disease heterozygous individuals can be shown to produce more offspring than those homozygous for the effective allele(s). The biological cause of this reproductive fitness is unknown. However, increased fitness is also observed in people heterozygous for sickle-cell anemia, who are more resistant than normal individuals to the effects of malaria.

Hemophilia A

Hemophilia A is a disease in which a normal level of the intrinsic blood-clotting Factor VIII is not present.

The gene for Factor VIII is located on the X chromosome. The hemophilic allele occurs at a frequency of one in ten thousand. Therefore the incidence of hemophilia is one in ten thousand for males but only one in one hundred million for females. Hemophiliacs suffer not only from external injuries but also from internal bleeding into their joints. This makes the disease painful and debilitating. Classical treatment includes transfusion of whole blood and injection of serum fractions rich in Factor VIII. As purer and more concentrated preparations of Factor VIII have become available, treatment procedures have improved. Unfortunately the cost of treatment is high because hemophiliacs may require large doses of purified material.

Biochemical analyses of afflicted individuals and their mothers (from whom they inherited the deleterious allele) have revealed a number of interesting facts about the genetic defect. Hemophiliacs have Factor VIII antigen, i.e., a protein which cross-reacts with antibodies which recognize purified Factor VIII. Their mothers exhibit clotting times anywhere from a few percent of the normal value to the normal value itself. This indicates a varied amount of Factor VIII. Whatever their clotting times, all carrier mothers exhibit a normal level of Factor VIII cross-reacting material (CRM).

Applications of Genetic Engineering Techniques to the Clinical Management of Hemophilia A

Replacement of the Missing Protein Factor

Hemophilia A represents a genetic disorder for which both current and future genetic engineering techniques will be of value. The popular press reports that one genetic engineering company has cloned the gene for human Factor VIII and will soon begin trials with it. This means that human DNA has been linked to the DNA of a microorganism and that a way has been found to make the microorganism transcribe and translate the human DNA. Thus the company should be able to procure human Factor VIII from large fermentations of the microorganisms. What remains to be seen is how well microorganism-produced material will work in humans.

Identification of Fetuses with the Disease

Carrier mothers generally come from families with a history of hemophilia and can be diagnosed by measurement of low blood clotting times. At present there are only two ways to prevent the occurrence of the disease in the progeny of such carriers: either the carriers can decide not to bear children, or they can have an abortion when amniocentesis reveals that the fetus is a male. But half of these male fetuses will be normal, making half of these abortions unwarranted. Currently used recombinant DNA procedures may soon be applied in such cases to distinguish between male fetuses with normal and those with mutant Factor VIII alleles, so that abortions would only be performed on afflicted fetuses. The methodology involves studying both the mutant and normal DNA (encoding the gene) to see if *easily* recognizible differences exist. For example, the DNA carrying the sickle-cell allele can be distinguished from normal DNA by a change in a restriction endonuclease site. Such a difference can be distinguished within the DNA of fetal cells from an analysis of cells in the amniotic fluid.

The ultimate contribution of genetic engineering techniques may be the ability to replace the defective DNA with cloned DNA from a normal individual.

Procedures

Students view the NOVA film *The Genetic Chance*, which covers Hemophilia A. This film may be rented or purchased from:

King Features Entertainment
235 East 45th Street
New York, New York 10017

Selected References

Motulsky, A. G. 1983. Impact of Genetic Manipulation on Society and Medicine. *Science* 219:135–140.

Stanbury, J. B., Wyngaarden, J. B., Fredrickson, D. S., Goldstein, J. L., and Brown, M. S., eds. 1983. *The Metabolic Basis of Inherited Disease*, 5th ed. New York, McGraw-Hill.

Problems

1. In approximately what fraction of families with five children will all of these children be of the same sex?

2. Two apparently normal parents produced three boys who died in infancy with similar symptoms of

disease. Two other children, a boy and a girl, appear to be normal. Which of three inheritance patterns under consideration—autosomal dominant, autosomal recessive, and sex-linked recessive—might be responsible for such a history?

3. Dominant defective alleles are rare but not unknown. Describe how a defective allele might show dominance in each of the following situations.

a. The gene codes for a membrane protein. The defective allele permits potassium ions to leak from the erythrocytes.

b. The gene codes for an enzyme composed of four identical polypeptide subunits.

4. What will limit the number of describable genetic diseases?

5. The film *The Genetic Chance* raises a number of questions that apply to the correlation between the mechanism of inheritance and the clinical observations. Try to think about these points as you watch the film. Hemophilic males have cross-reacting material (CRM) to antibodies directed against Factor VIII. The antigen is not Factor VIII itself but another factor which is tightly bound to Factor VIII in the serum. Thus, it is difficult to determine whether an abnormal Factor VIII protein is present. Nevertheless, there is a range in the degree of severity of the disease. What does this mean in terms of the types of mutations which give rise to hemophilia A? With the knowledge that there is a large range in severity and that carrier mothers can exhibit an extremely high clotting time, speculate on the amount of functional Factor VIII which is necessary. Do you believe a complete absence of Factor VIII could be tolerated? If you could sequence the structural gene for Factor VIII from different individuals, what would you expect to find?

6. Another question raised by the film concerns carrier mothers who exhibit a large range in clotting times. Why should this be so?

7. Based on your knowledge of the structure and biosynthesis of proteins, how might the microbial product of a cloned human gene differ from the normal human product? What is the clinical significance of this difference?

8 Use of Radioisotopes in Clinical Biochemistry

Ronald D. Edstrom and Robert P. Chandler

Biochemical processes such as hormone-cell interactions, gene replication, and protein synthesis take place in dilute solutions or at low concentrations at membrane surfaces. These low concentrations require extraordinarily sensitive methods of detection and measurement. The high sensitivity of instruments used in detecting radioactivity permits the performance of experiments that would fail if ordinary chemical methods were used. Chemical methods of analysis are generally sensitive to a few nanomoles at best (10^{-9} mole). By using radioisotopes, molecules can be detected in the femto- (10^{-15}) to picomole (10^{-12}) range.

Many metabolic pathways have been elucidated with the aid of ^{14}C and ^{3}H. The pentose phosphate pathway of aerobic glucose metabolism was discovered with the use of ^{14}C-labeled glucose and other sugars in which specific carbons in the sugar backbone were labeled with the radioactive marker. The role of CO_2 in the biosynthesis of fatty acids was discovered by using ^{14}C-labeled bicarbonate. Thousands of metabolic interconversions have been described using radioisotopes to track specific atoms in complex molecules. Nearly all of our understanding of protein synthesis and nucleic acid synthesis is based on radioisotope data.

Many of the radioisotope techniques employed in biomedical research have been adapted for use as clinical testing methods. Examples are determination of body fluid volume during treatment of severe burns and analysis of hormone levels in endocrinologic disorders. Clinicians utilize ionizing radiation from isotopes or accelerators for diagnosis (X rays) or for treatment of many malignant diseases.

The radioactive isotope (radionuclide) most commonly used in the Nuclear Medicine Clinic at the University of Minnesota is technetium−99m. The purpose of today's demonstration is to familiarize students with some properties of this radionuclide. The physical half-life of the technetium−99m will be determined, the half-value layer (the amount of shielding that reduces the radiation intensity to one-half) will be measured, and the effect of distance from the source of technetium−99m on the radiation intensity will be observed.

Principles

Definition of an Isotope

The word *isotope* is Greek, meaning "same place". Webster's "*New Collegiate Dictionary*" defines it as "any of two or more species of atoms of a chemical element with the same atomic number and position in the periodic table and nearly identical chemical behavior but with differing atomic mass". Every molecule, natural and unnatural, is constructed exclusively of isotopes. All the elements of the periodic table exist as sets of isotopes. For example, there are three for hydrogen, six for carbon, seven for nitrogen, and seven for oxygen. There are about 2,000 known isotopes of the 106 elements. Table 8.2 (see Appendix) gives the isotopes of hydrogen, carbon, nitrogen, and oxygen. It follows the convention of making the atomic number a subscript and the atomic mass a superscript, both to the left of the symbol for the element. Thus oxygen, nitrogen, and carbon all have isotopes with a mass of 15 but differ in their atomic numbers.

Isotopes are either stable or unstable. Stable isotopes will not change their nuclear composition in the absence of an outside influence. In contrast, unstable isotopes will spontaneously undergo a nuclear conversion to a different element, with the ejection of energy from the nucleus in the process, called radioactive decay. About 1,600 of the 2,000 isotopes are unstable.

Principles of Radioactive Decay

The nuclei of unstable isotopes contain a proportion of neutrons and protons that will ultimately disintegrate and leave a more stable nucleus. Thus ^1_1H is a stable isotope and ^2_1H is stable, but with the addition of a second neutron to form ^3_1H the nucleus becomes unstable and emits an electron with the concomitant conversion of one of the neutrons into a proton:

$$^3_1H \rightarrow {}^3_2He + e^-.$$

1 proton	2 protons
2 neutrons	1 neutrons

The product of the reaction is no longer an isotope of hydrogen but rather one of helium with an atomic number of 2 and mass of 3. *The electron expelled during such a nuclear conversion is called a beta- (β-) particle. It carries the negative charge and mass of an ordinary electron but is ejected with a substantial kinetic energy, which allows it to penetrate matter and ionize atoms with which it might collide.*

Some unstable isotopes of high atomic mass transmutate by the elimination of a large piece of matter, *the alpha- (α-)particle, which consists of 2 protons and 2 neutrons. The α-particle is simply a helium nucleus* – $^4_2He^{+2}$. Radium – 226 is an α-emitter.

$$^{226}_{88}Ra \rightarrow {}^{222}_{86}Rn + {}^4_2He^{+2}$$

88p, 138n	86p, 136n	2p, 2n

The ejected α-particle carries substantial kinetic energy and with its +2 charge is highly ionizing when it interacts with matter.

The third type of nuclear change for an unstable isotope, gamma- (γ-)decay, is substantially different from the previous processes. For both α- and β-decay, the primary event is the transmutation of the nucleus by emission of a charged particle with substantial kinetic energy. In one type of γ-decay, the nucleus may be changed by its "capturing" an electron from the atomic orbitals. This process, called electron capture, results in the conversion of a proton to a neutron. Invariably energy is produced in this conversion and emitted as electromagnetic radiation in the form of a γ-ray. *γ-radiation is not particulate except in the sense of the duality in the wave-particle nature of electromagnetic radiation.* γ-radiation has the *energy characteristics of X rays and is quite penetrating and ionizing*. Iodine – 125 is a typical, electron capture γ-emitter.

$$^{125}_{53}I + e^- \rightarrow {}^{125}_{52}Te + \gamma$$

53p, 72n	52p, 73n

In addition to electron capture, γ-radiation may be produced subsequent to an α- or β-particle emission in which the nucleus is in an excited or metastable state. The excess energy is then emitted as γ-ray as the nucleus returns to the ground state. The decay of molybdenum – 99 to form technetium – 99m is an example of the decay of a metastable nucleus.

$$^{99}_{42}Mo \rightarrow \beta + {}^{99m}_{43}Tc \quad t\frac{1}{2} = 3 \text{ days}$$

$$^{99m}_{43}Tc \rightarrow {}^{99}_{43}Tc + \gamma \quad t\frac{1}{2} = 6 \text{ hours}$$

Technetium – 99m has an "excited nucleus" which emits energy as a γ-ray to reach the lower-energy ground state.

Detection Methods

The detection of the radiation emitted by a decaying nucleus is based on the interaction of the emitted particle or photon with a gas, liquid, or solid substance in the detector. The nature of this interaction depends on both the type of emission and its energy. Each emitted particle or photon has energy resulting from the loss of mass during the decay process. Ions are produced when emitted radiation interacts with matter; the ions are measured directly by means of an ion chamber, photographic film, or scintillation counting.

Exposure of a gas (helium, argon) to radiation in an electric field causes ionizations and results in an electric current. The current may be measured as an indication of the intensity of radiation, or the individual ionizing events may be counted and reported as counts per minute (CPM). Ion chambers are most efficient for α- and β-radiation, for these two types of particles interact strongly with low-density matter in a short distance (Figure 8.1).

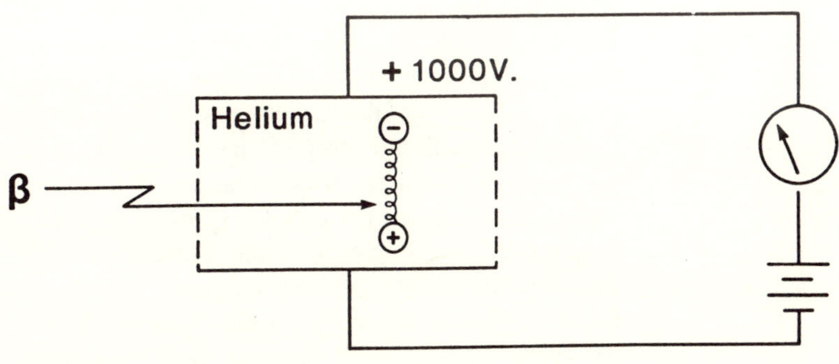

Figure 8.1. Ion chamber detector

A major problem of *ionization* detectors of this type is the difficulty of enclosing the ion chamber to keep the gas in, while still allowing α- and weak β-particles to enter. The isotope is introduced into the chamber in some cases, or an open chamber can be used with a continuous flow of gas. For γ-rays and high energy β- (>0.1 MeV) emission, an ion chamber with a thin window of mylar can be used. Gaseous ionization chambers are generally used only as safety monitors and as highly specialized detectors in biological research.

The use of *photographic film* for detection of ionizing radiation has been the basis of the X ray for clinical diagnosis. (X rays are identical to low-energy γ-radiation.) In research laboratories photographic film is used for safety monitoring (film badges) and as a means of locating radioisotopes in two dimensional assays—paper chromatography, gel electrophoresis, and so on. The silver halide crystals in photographic film are ionized by the radiation, and when the film is developed with a reducing agent, a darkened area of metallic silver is produced. The density of the darkened area is proportional to the radiation dose.

The most widely used method of quantitation of radioisotopes in biological science is *liquid scintillation counting*. When a β-emitter is mixed with a solution of any one of several different organic molecules, called fluors, the ionization of the organic substance by a β-particle produces a flash of light. Efficiencies greater than 80% can be achieved with this method (Figure 8.2).

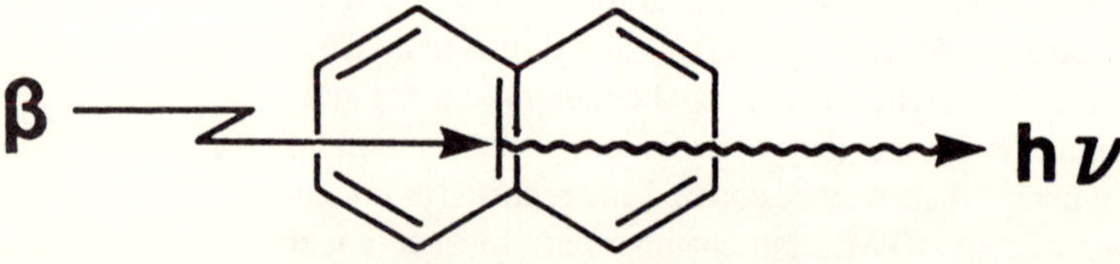

Figure 8.2. Scintillation fluor (naphthalene) produces light when struck by ionizing radiation

Naphthalene is a fluor that yields up to 30 photons for each β-particle collision; the light produced is measured with an electronic photodetector. The number of photons emitted per α- or β-particle is dependent on the energy of the particle; therefore the intensity of the light flash is proportional to the energy of the radiation. The detector quantitates both the number of ionizing events per minute expressed as counts per minute and the energy of the ionizing particle. Thus it is possible to quantify simultaneously two different isotopes in the same mixture, such as ^{14}C and ^{3}H. When high-energy β-particles (> 1 MeV) pass through a relatively dense medium, such as water, energy is given off as light (Cerenkov) radiation, which can be quantitated. The biologically important isotope ^{32}P is often measured in this way.

For γ-ray quantitation it is more effective to use a *solid scintillation counter*. A crystal of fluor material is subjected to the γ-radiation from the isotope in a glass or plastic tube. Since the γ-radiation easily passes through the tube, one need not mix the fluor with the isotope.

Safety in Radioisotope Use

While no ionizing radiation is harmless, the experiments in this manual have been designed to limit the hazards from radiation to *below* the level of normal life risks (walking across the street, driving to work, and so on). The radiation from the radioactive materials used are orders of magnitude smaller than that received from one diagnostic X ray.

Nevertheless, a few rules must be followed. The best general precaution to use when working with isotopes is to avoid exposure.

1. Keep isotopes out of your body and off your skin.
2. Avoid exposure to ionizing radiation by maintaining a proper distance or by using a shield.

Never pipette radioactive materials by mouth, even with rubber tubes. Use of protective clothing (gloves, etc.) may be appropriate in some cases, but only gloves are needed for the experiments in this course.

External sources emit radiation in all directions; thus the exposure rate decreases as the square of the distance increases. For large amounts of isotopes, extra shielding may be required.

α-particles will not penetrate skin and will be stopped in the first four inches of air.

β-particles from low-energy sources like ^{3}H will not escape from ordinary laboratory glassware. For high-energy emitters like ^{32}P, a shield made of 1/2-inch lucite is usually used.

γ-rays are attenuated by heavy metal such as lead. The thickness needed can vary from a few mm(^{125}I) to several inches (^{60}Co).

In the experiments you will perform, no shielding is needed.

Clinical Applications

Early uses of radioisotopes in the practice of medicine were primarily oriented toward treatments, as in the introduction of small pieces of radium into solid tumors for cancer therapy. At present nearly all (>95%) of the medical uses of radioactive isotopes are diagnostic.

The same kinds of uses mentioned for research are adaptable to clinical *in vitro* tests, e.g., enzyme assay, radioimmunoassay. Several uses of technetium–99m and xenon–133, two of the most widely used isotopes for *in vivo* diagnostic techniques, are listed in Table 8.1. For *in vivo* techniques, low-energy, short-lived γ-emitters are used. The short half-life (a few days at most) ensures minimum exposure to the patient. The use of a γ-emitter allows an externally located crystal scintillation detector to locate and quantify the isotope. Neither α- nor β-emission could be used because these particles would be absorbed before reaching the surface of the body.

Table 8.1. Diagnostic Uses of Technetium-99m and Xenon-133

Isotope Used	Method of Introduction	Physiological Parameter Measured
Technetium-99m	Pertechnetate salt	Brain transit time
	Albumin complex	Blood volume, blood flow
	Microspheres (encapsulated)	Lung blood flow, gastric emptying, peripheral thrombi
Xenon-133	Inhalation	Pulmonary ventilation, cerebral blood flow
	Intravenous (saline solution)	L→R cardiac shunt, pulmonary blood flow, cardiac output (right heart)
	Intraarterial (saline solution)	Renal blood flow, L→R blood flow
	Local (saline solution)	Extremity blood flow

Two common examples illustrate the utility of γ-emitters for clinical diagnosis. The first involves injection of technetium–99m-labeled serum albumin microspheres (SAM) to detect pulmonary emboli. The technetium–99m SAM is injected intravenously, and the microspheres are trapped in the capillaries of the lungs, which have not been shut off by an embolism. By using a rectilinear crystal γ-ray scanner, an image of the lung vasculation shows nonradioactive areas not receiving blood as a result of the embolus.

A second example is ventilation scintiphotography of the lungs after inhalation of xenon–133. It is possible to evaluate the distribution of inhaled gases in the various areas of the lungs as well as the rate of washout for a specific portion of the lung that may have impaired ventilation.

As with any invasive medical procedure, the physician must evaluate the risk-benefit probabilities carefully.

Procedures

Half-Life

With a G-M survey meter and pancake probe, determine the count rate of a technetium–99m source and record the results. Repeat the procedure periodically during the demonstration. Record your results (see Sample Worksheet 8.1). On semilog paper, plot the activity of the source on the y-axis against elapsed time on the x-axis. Draw the best straight line to connect the points. From this graph, compare the measured half-life of technetium–99m with the handbook value of 6.05 hours.

Sample Worksheet 8.1. Determination of the Half-Life of Technetium-99m

Date	Time	Activity of Source

Half-Value Layer (HVL)

Place a G-M survey meter and pancake probe over a source of technetium–99m and determine the count rate. Then place a 0.711 mm thick sheet of lead between the source and the probe. Note the count rate and record on a worksheet (see Sample Worksheet 8.2). Repeat with additional 0.711 mm sheets of lead until four sheets have been used. Plot the results on semilog paper, with the count rate on the y-axis and absorber thickness on the x-axis. Determine the half-value layer graphically and compare with the handbook value of 0.3 mm of lead HVL for technetium–99m.

Sample Worksheet 8.2. Determination of the Half-Value Layer

Absorber Thickness (mm)	Count Rate (cpm)
None	
0.711	
1.422	
2.133	
2.844	

Inverse Square Law

Place a technetium−99m source against a meter stick and place a pancake probe attached to a G-M survey meter at 10 cm. Note the count rate and record on a worksheet (see Sample Worksheet 8.3). Repeat for 20 cm, 40 cm, and 80 cm. Plot the results on linear graph paper, with the count rate on the y-axis and $1/d^2$ on the x-axis. Fit the best straight line.

Sample Worksheet 8.3. Determination of the Inverse Square Law

Distance (cm)	Distance^2 (d^2)	$1/d^2$	Count Rate (cpm)
10	100	0.01	
20	400	2.5×10^{-3}	
40	1600	6.25×10^{-4}	
80	6400	1.56×10^{-4}	

Selected References

Beebe, G. W. 1982. Ionizing Radiation and Health. *American Scientist* 70:35–44.

Cerny, J., and Poskanzer, A. M. 1978. Exotic Light Nuclei. *Scientific American* 238:60–72.

Loken, M. K., and Payne, J. T. 1975. Benefit vs Risk in the Practice of Nuclear Medicine. *Minnesota Medicine* 58:173–186.

Problems

1. Write the decay equation for (a) carbon−14 and (b) phosphorus−32.

2. What would be an appropriate shield for protection against the radiation of (a) iodine—125, (b) phosphorus—32, (c) tritium, and (d) technetium—99m?

3. How long would it take for 1 mCi of technetium−99m to decay to below 1 μCi?

4. What method of detection would be appropriate for iodine—125, phosphorus—32, tritium, uranium—235, and technetium—99m?

Appendix

Table 8.2. Radioactive and Stable Isotopes of Hydrogen, Carbon, Nitrogen, and Oxygen

Element	Isotopes
Hydrogen	${}^{1}_{1}H$, ${}^{2}_{1}H$, ${}^{3}_{1}\underline{H}$
Carbon	${}^{10}_{6}\underline{C}$, ${}^{11}_{6}\underline{C}$, ${}^{12}_{6}C$, ${}^{13}_{6}C$, ${}^{14}_{6}\underline{C}$, ${}^{15}_{6}\underline{C}$
Nitrogen	${}^{12}_{7}\underline{N}$, ${}^{13}_{7}\underline{N}$, ${}^{14}_{7}N$, ${}^{15}_{7}N$, ${}^{16}_{7}\underline{N}$, ${}^{17}_{7}\underline{N}$
Oxygen	${}^{14}_{8}\underline{O}$, ${}^{15}_{8}\underline{O}$, ${}^{16}_{8}O$, ${}^{17}_{8}O$, ${}^{18}_{8}\underline{O}$, ${}^{19}_{8}\underline{O}$, ${}^{20}_{8}\underline{O}$

Underlined elements are radioactive. All the radioactive isotopes of nitrogen and oxygen have such short half-lives that they are of almost no use in biological research. Of the carbon radioisotopes, only ^{14}C and ^{11}C are useful, although the short half-life (20.5 min) of ^{11}C limits its usefulness in most situations.

Energy Produced in Radioactive Decay

In all three types of radioactive decay, the energy of the emitted particle, or quantum of radiation, results from the conversion of mass into energy. The mass of the products is less than that of the starting materials. The lost mass is converted into energy which can be numerically evaluated by Einstein's equation:

$$E = mc^2$$

where m = mass in grams, c = speed of light (3×10^{10} cm/sec), E = ergs.

The unit used in describing the energy of radioactive emissions is the electron volt (eV), which is the amount of kinetic energy acquired by an electron in passing through a potential gradient of 1 volt. The magnitude of the energy of radioactive emissions ranges from a few thousand to several million electron volts (1 electron volt is equal to 1.6×10^{-12} erg). The energies of the emitted radiation for several isotopes are listed in Table 8.3.

The α-particles emitted in a reaction such as the disintegration of radium−226 will each have an energy of 4.8 MeV. This kinetic energy corresponds to a velocity of about 2×10^9 cm/sec. The combination of high charge density and large mass of the α-particles results in a high frequency of interaction of the particle with other atoms so that the range is only 2.6 cm in air and the particles will be stopped completely by a sheet of paper or the skin of an animal. However, along that brief track of interaction the amount of ionization is quite high.

The energies indicated for β-emissions are maximal values for the isotopes listed. Unlike α-particles, β-particles (electrons) have a distribution of energies, with the average β-particle having substantially less than the maximal energy. Typical spectra of the energies for β-particles are shown in Figure 8.3. Since each disintegration of an atom of a specific isotope results in the release of the same amount of energy, those events that produced β-particles of less than the maximum energy must dissipate the energy differences in another form. This is accomplished by the production of a neutrino, an uncharged particle with a kinetic energy equal to the difference between the actual energy of the β-particle and the maximal energy E_{max} for a β-emission from the specific atom. Since neutrinos can travel many miles without interacting with matter, they are considered to be neither useful nor hazardous in biological systems. In contrast, the negatively charged β-particles have a high probability of interacting with matter. Their range is dependent on the density of the medium and energy of the emission. In dry air the range is about four meters per MeV. It takes an energy of 0.07 MeV to penetrate human skin.

The third form of ionizing radiation to be considered, γ-rays, is nonparticulate and emitted with discrete energy levels. Like other electromagnetic radiations, they can be attenuated by being absorbed though interaction with matter. High atomic weight materials such as lead are the most effective absorbers. The range of γ rays in dry air is four meters per MeV. They will easily penetrate the skin if they have energies greater than 0.07 MeV.

A major consideration in working with radioisotopes is their rate of decay. Each unstable isotope has a characteristic probability of disintegration. It is a statistically predictable event only when dealing with large numbers of atoms. A convenient tradition for describing that probability is the use of the concept of

Table 8.3. Characteristics of Radioisotopes Commonly Used in Biology and Medicine

Radioisotope			
Symbol	Name	Energy and Emission Type	Half-Life
$^{3}_{1}H$	Tritium	0.019 MeV β	12.5 years
$^{14}_{6}C$	Carbon-14	0.16 MeV β	5,720 years
$^{11}_{6}C$	Carbon-11	0.96 MeV β, Mev γ	20.5 minutes
$^{32}_{15}P$	Phosphorus-32	1.7 MeV β	14.3 days
$^{45}_{20}Ca$	Calcium-45	0.25 MeV β	164 days
$^{99m}_{43}Tc$	Technetium-99m	0.14 MeV γ	6 hours
$^{125}_{53}I$	Iodine-125	0.035 MeV γ	60 days
$^{131}_{53}I$	Iodine-131	0.6 MeV β, 0.364 Mev γ	8 days
$^{133}_{54}Xe$	Xenon-133	0.35 MeV β, 0.08 Mev γ	5 days
$^{137}_{55}Cs$	Cesium-137	0.52 MeV β, 0.66 Mev γ	33 years
$^{226}_{88}Ra$	Radium-226	4.8 MeV α, 0.18 Mev γ	1,600 years
$^{238}_{92}U$	Uranium-238	4.2 MeV α, 0.05 Mev γ	4.5×10^9 years
$^{239}_{94}Pu$	Plutonium-239	5.1 MeV α, 0.04 Mev γ (other γ)	24.4 years

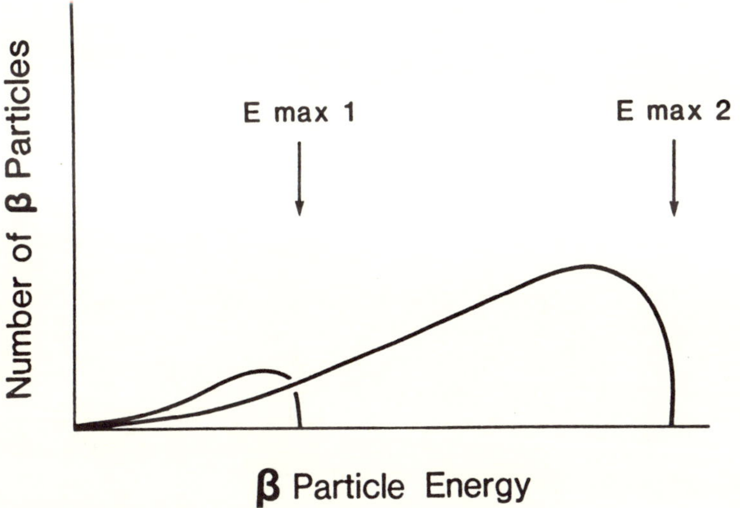

Figure 8.3. Distribution of energies for the β-particles of two different isotopes

half-life ($T_1/_2$). The half-life is the amount of time necessary to allow half of the atoms to decay. The range of known half-lives is from less than a microsecond (3×10^{-7} sec for ^{212}Po) to billions of years (4.5×10^9 yr for ^{238}U). The half-life is independent of concentration or amount of isotope or any other physical influence such as temperature, pressure, and so on. The fraction of isotope left after n half-lives is equal to $\frac{1}{2^n}$. Notice that no matter how large a finite number n becomes, the fraction never reaches zero. Figure 8.4 shows decay curves for tritium. The same data are expressed on a linear basis (left) and on a logarithmic scale (right).

Definitions

Curie. Quantity of radioactive material that will give 3.7×10^{10} disintegrations per second.

Electron-Volt. Energy imparted to an electron while passing through a one volt potential gradient.

Half-Life. Time required for a radioactive substance to lose one-half of its activity by decay.

Rad. Amount of absorbed radiation equivalent to 100 ergs per gram of absorbing material.

Rem. Special unit of dose equivalence. Rem = rads × quality factor.

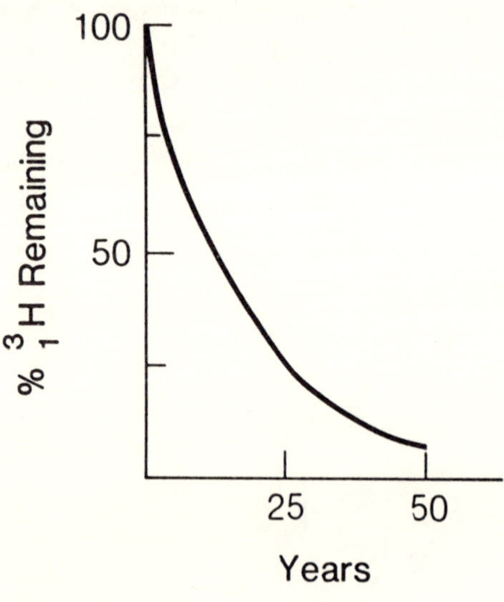

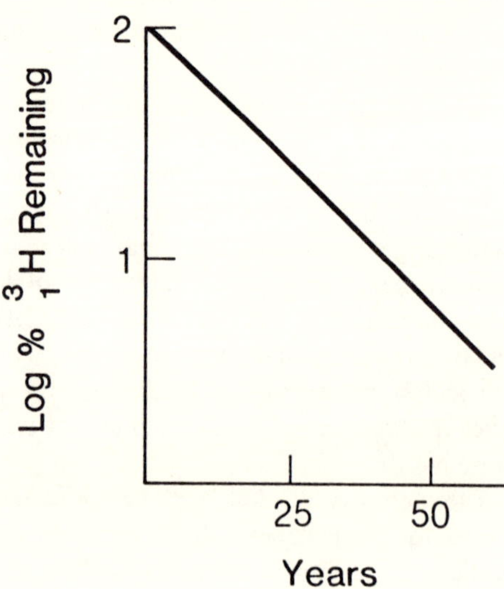

Figure 8.4. Decay curves of tritium (half-life 12.5 years)

Description of the Atom

The *atom* is the smallest subunit of matter to which a chemical property can be ascribed. An atom consists of a nucleus surrounded by electrons. The nucleus comprises most of the mass of the atom and with the exception of hydrogen (mass = 1), nuclei contain both the positively charged protons and the electrically neutral neutrons. The weights of these two major nuclear particles are nearly the same and are each assigned a mass of 1. The electrons in the various orbitals of the atom outside the nucleus are negatively charged particles with a mass 1/1838 that of a neutron or proton. The electrons in their orbitals provide the chemical properties of the atom. In a nonionized atom the total number of electrons in all the orbitals is equal to the number of protons in the nucleus. That number is called the *atomic number*. The sum of the number of protons and neutrons is referred to as the *atomic mass* or weight.

Roentgen. Amount of X rays or γ-rays required to produce ions carrying one electrostatic unit of charge in dry air under standard conditons.

Common Examples of Radiation Exposure

Background	Typical U.S. resident—0.1 rem/yr (double at Denver and other high elevations due to increase in cosmic-ray intensity)
	Commercial airline pilot—0.16 rem/yr
	Worker in nuclear power plant—0.4 rem/yr
Radiation from diagnostic X rays	Chest X ray—0.01 rad Lower GI barium contrast series—0.9 rad

A Few Subatomic Particles

Name	*Mass*[a]	*Charge*
Electron	$\frac{1}{1838} = (0.000544)$	−1
Neutron	1	0
Neutrino	0 (at rest)	0
Positron	$\frac{1}{1838} = (0.000544)$	+1
Proton	1	+1
α-Particle	4	+2

[a]Given as a fraction of the mass of a hydrogen atom.

9 Determination of Glycosylated Hemoglobin

Marilyn H. Koenst and Ronald D. Edstrom

Approximately 90% of normal adult human hemoglobin is the $\alpha_2\beta_2$ tetramer called hemoglobin A. Four to 7% of hemoglobin A is present as glycosylated hemoglobin, which contains covalently-bound glucose and can be distinguished from hemoglobin A by various methods. In persons with high blood-glucose levels the amount of the glycosylated forms of hemoglobin also will be high, and thus the measurement of glycosylated hemoglobin is of clinical importance. In this experiment we will measure the relative amounts of glycosylated hemoglobin in blood from normal and diabetic individuals.

Principles

Nonenzymatic Formation of Glycosylated Hemoglobin

The hemoglobin of normal persons contains a few percent of molecules that have a relatively low isoelectric point and therefore move more rapidly than most of the hemoglobin upon electrophoresis and cation exchange chromatography. This so-called fast hemoglobin is the result of the nonenzymatic addition of glucose to the amino-groups present on some of the lysine residues within the peptide chains and N-termini of the hemoglobin molecules.

The erythrocyte membrane is freely permeable to glucose, so the hemoglobin in the red cells is constantly exposed to glucose at the same concentration as it is in the plasma. In the presence of free amino groups, aldoses form N-glycosides that can undergo an irreversible Amadori rearrangement to form a stable glycosyl derivative (Figure 9.1).

Methods for Determining Glycosylated Hemoglobin

Two methods have been used for the measurement of glycosylated hemoglobin: (1) colorimetric analysis of the carbohydrate derivatives; (2) separation and subsequent quantitation of the modified and unmodified proteins.

All the colorimetric methods depend on the acid hydrolysis and dehydration of the carbohydrate residue to form 5-hydroxymethylfurfural, which can then be determined with the reagent 2-thiobarbituric acid. A yellow color, formed as the result of the reaction of 5-hydroxymethylfurfural with 2-thiobarbituric acid, can be evaluated spectrophotometrically (Figure 9.2). The colorimetric methods have the advantage of using simple reagents; however, these methods are subject to substantial error through small changes in the reaction conditions. Additionally, the lack of appropriate standard substances makes it difficult to determine the absolute extent of glycosylation of the hemoglobin sample.

Separation of the glycosylated and unmodified hemoglobin can be achieved by a variety of techniques. Once separated, the amounts of the two forms of hemoglobin can be determined spectrophotometrically by measuring the absorption at 415 nm. Ion exchange chromatography with cation exchange resin is the basis of several useful separation methods. The modification of a single amino group on a large proteinlike hemoglobin does not make a very large change in the ionic properties. As a consequence, the separation of the two forms by ion exchange chromatography is quite sensitive to small changes in pH and temperature. That is, the pH must be maintained with a variation of less than 0.02 units and temperature

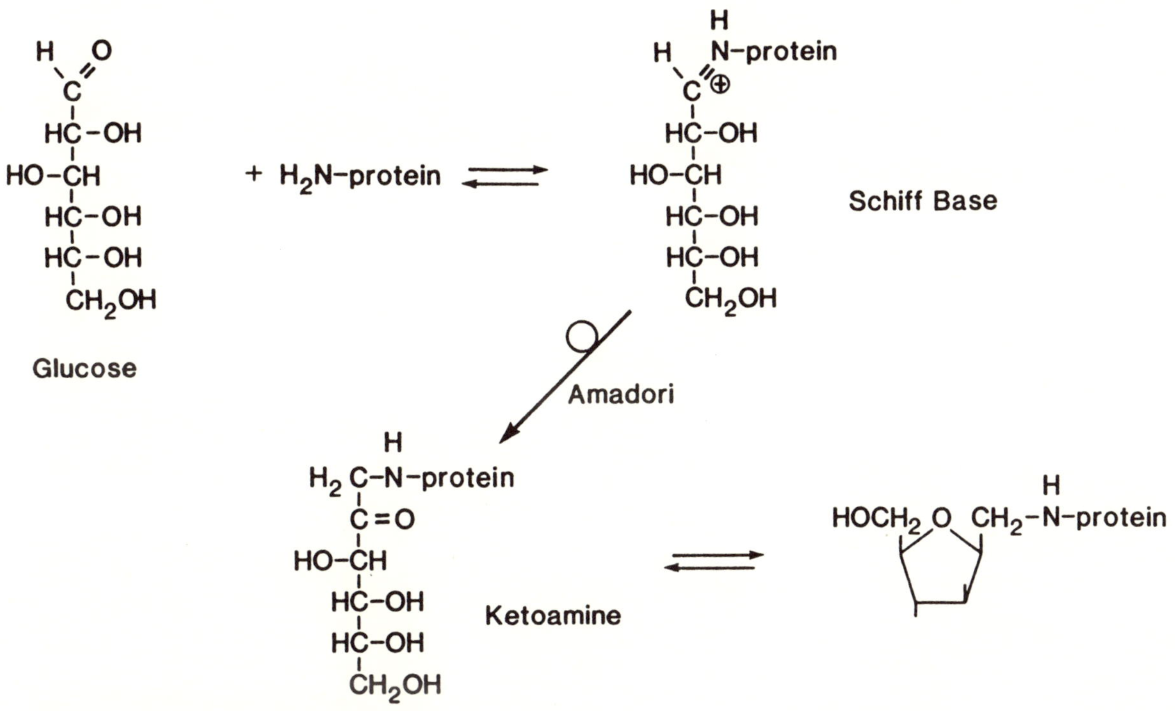

Figure 9.1. Nonenzymatic reaction of glucose with protein

within 1° or 2°C. Such close control of pH and temperature is difficult. Electrophoresis and isoelectric focusing are techniques that can be used to separate the two types of hemoglobin; however, they too are sensitive to inaccuracies due to slight variations in conditions of the procedures. Additionally, these two techniques usually require densitometric measurements of the hemoglobin bands in the gel or other matrix, and densitometric scanning is relatively inaccurate compared to colorimetry.

The simplest way to separate the glycosylated from unmodified hemoglobin is by use of affinity chromatography. Compounds such as carbohydrates will form cyclic esters with borate compounds (Figure 9.3). If the borate compound is covalently attached to an insoluble matrix, a column can be prepared that will selectively adsorb substances with two adjacent hydroxyl groups. Glycosylated hemoglobin will adhere to such a column, whereas the unmodified hemoglobin will pass through. After washing the column, the bound glycosylated protein can be eluted by passing a relatively concentrated solution of a compound such as 0.2 M sorbitol (glucitol) through the column. The polyhydric alcohol at high concentrations forms esters with the borate and displaces the glycosylated hemoglobin, which cannot effectively compete at its much lower concentration, on the order of 10 μM. The concentration of the hemoglobin in the two fractions (wash and eluate) is then determined spectrophotometrically, and the percentage of glycosylated hemoglobin can then be calculated.

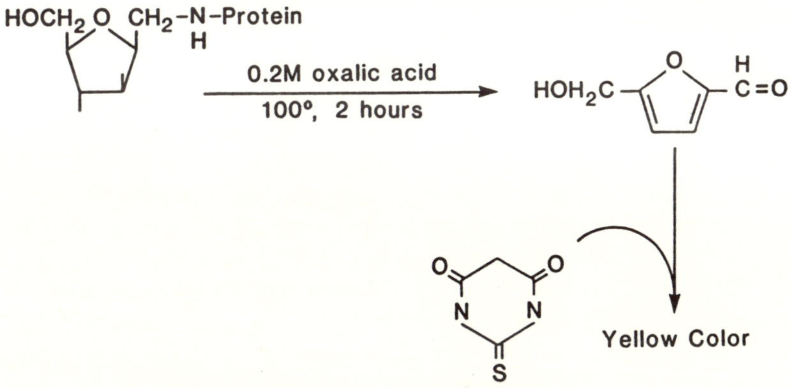

Figure 9.2. Colorimetric determination of glycosylated hemoglobin

Clinical Applications

Individuals with uncontrolled diabetes have high levels of blood glucose. This increases the rate of non-

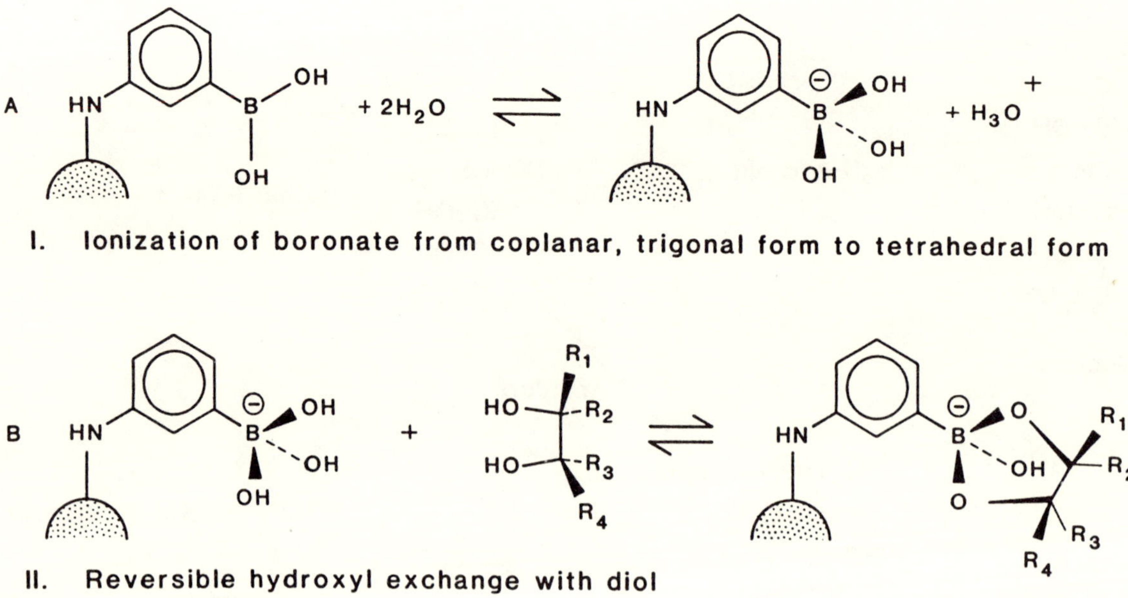

I. Ionization of boronate from coplanar, trigonal form to tetrahedral form

II. Reversible hydroxyl exchange with diol

Figure 9.3. Affinity chromatography of glycosylated hemoglobin

enzymatic glycosylation of hemoglobin. Persons with normal blood glucose levels have from 4 to 7% of their hemoglobin in the glycosylated form, whereas in poorly controlled diabetics the value may be as high as 18 to 20%.

Successful clinical management of blood-sugar levels in diabetics depends on the physician's ability to evaluate blood-glucose concentrations over long periods. Even though techniques for patient self-monitoring are readily available and generally reliable, it is important for the physician to be able to make an independent verification of the level of control. That is especially true for patients who cannot or will not regularly follow and accurately report on the self-monitoring procedure.

An increasingly important technique for evaluating long-term trends in blood-glucose control is the measurement of glycosylated hemoglobin. Once formed, the modified hemoglobin molecules stay in the circulation for the life of the erythrocyte (120 days). Thus, whereas an ordinary blood-glucose determination gives an indication of blood-sugar control at the moment the blood was drawn, measurement of glycosylated hemoglobin reveals an integral value for blood-glucose levels for several weeks before the determination.

Procedures

Prepacked columns of m-amino phenylboronic acid attached to an insoluble matrix (Sepharose CL−6B) are used (see Appendix Reagents, p. 103). The columns are contained in polyethylene tubes with a small built-in reservoir at the top. The columns are inserted into plastic caps that will support them over the test tubes used to collect the fractions. Two different buffer solutions are used in this experiment. The wash buffer is in polyethylene squeeze bottles, whereas the eluting buffer (containing sorbitol) is in a glass container. Each group of four students determines glycosylated hemoglobin values for their own blood and for a control specimen (diabetic blood).

1. Equilibrate the resins by allowing about 5 ml of wash buffer (0.25 M ammonium acetate, pH 7.8, containing 0.02% NaN_3 and 1% Triton X−100) to pass through the columns. Discard the effluent.

2. Mix the hemolysate samples and dilute each 1:2 with wash buffer.

3. Fit the columns into graduated 20 × 200 mm test tubes. The nonglycosylated hemoglobins are collected in these tubes. Carefully pipette 100 μl diluted hemolysate onto a column; allow the sample to flow into the resin bed and then wash the sample into the column with about 0.5 ml of wash buffer. When the sample has been washed in, fill the tube and reservoir with wash buffer. Continue adding buffer and collecting the washings until about 15 ml of buffer have been collected.

4. Transfer the columns to graduated 16 × 150 mm tubes (these are calibrated to contain 10 ml). Fill the tube and reservoir with eluting buffer (0.1 M tris, pH 8.5 containing 0.2 M sorbitol, 0.02% NaN_3 and 1% Triton X−100) and collect the effluent which contains the glycosylated hemoglobins.

5. Bring the volume of the "wash" fractions (non-glycosylated hemoglobin) to 50 ml with wash buffer. Bring the volume of the "eluant" fractions (glycosylated hemoglobin) to 10 ml with eluting buffer. Mix the tubes well and transfer 3.0 ml samples of the contents to 3 ml spectrophotometer cuvettes. Measure the absorbance of each at 415 nm against a deionized water blank.

6. Calculate the percentage of glycosylated hemoglobin:

$$\%\text{ of glycosolated hemoglobin} = \frac{10(A_{415}\text{ eluant})}{50(A_{415}\text{ wash}) + 10(A_{415}\text{ eluant})} \times 100.$$

Selected References

Bunn, H. F. 1981. Evaluation of Glycosylated Hemoglobin in Diabetic Patients. *Diabetes* 30:613–617.

Bunn, H. F., Gabbay, K. H., and Gallop, P. M. 1978. The Glycosylation of Hemoglobin: Relevance to Diabetes Mellitus. *Science* 200:21–27.

Mallia, A. K., Hermanson, G. T., Krohn, R. I., Fujimoto, E. K., and Smith, P. K. 1981. Preparation and Use of a Boronic Acid Affinity Support for Separation and Quantitation of Glycosylated Hemoglobins. *Anal. Lett.* 14:649–661.

McFarland, K. F. 1981. Glycosylated Hemoglobin: What Is Its Value? *Arch. Int. Med.* 141:712.

Problems

1. Record the percentage of glycosylated hemoglobin in the samples you have analyzed. Also list the blood-glucose values obtained by the automated glucose-oxidase procedure.

2. Galactose is known to form glycosylated derivatives of hemoglobin at a faster rate than glucose. What might be the clinical implications in a patient suffering from galactosemia?

3. All blood proteins with free amino groups are glycosylated nonenzymatically by glucose. Why is hemoglobin a more useful indication of long-term hyperglycemia than serum albumin?

4. In an oral glucose tolerance test a large amount of glucose (100 g) is swallowed and the level of blood glucose determined at intervals of 30 minutes, one, two and three hours. Three glucose tolerance curves are shown in figure 9.4. Curve A depicts the normal glycemic response to the oral administration of 100 g of glucose. The rise in the blood-sugar level is rapid, but a normal value is restored in two hours. In curve B the glycemic response is slower, and normal values are not restored until the third hour, as is found in mild diabetes. Curve C depicts the fasting hyperglycemia and the continued increase in the blood-sugar level even at the third hour, as seen in severe diabetes. Glycosuria usually occurs when the blood sugar level is maintained (for several hours) above 160 mg per 100cc.

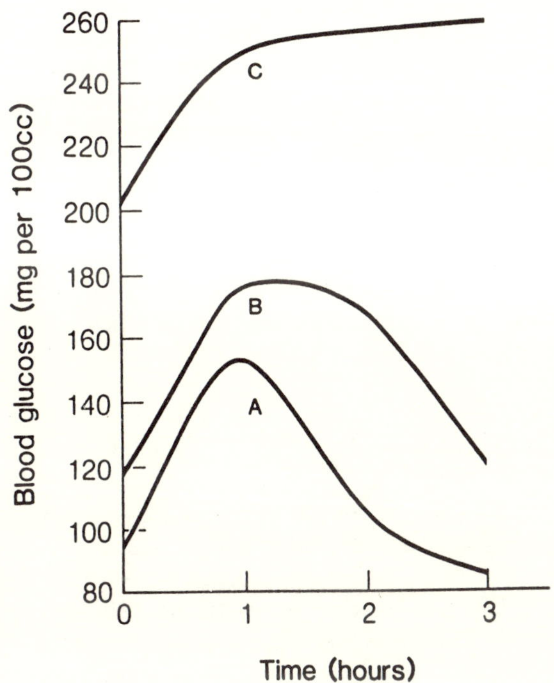

Figure 9.4. Three glucose tolerance curves

What would be the level of glycosylated hemoglobin in individuals who exhibit each of these responses (A, B, and C)? Would the levels of glycosylated hemoglobin of each individual change in the three hours after the ingestion of 100 g of glucose?

10 Biosynthesis of Adrenal Steroid Hormones

Frank Ungar

The adrenal cortex converts cholesterol to biologically active steroid hormones. More than 30 steroid compounds have been isolated from adrenal tissue. The analysis of adrenal steroids in blood and urine is used in the diagnosis of a variety of disease states, e.g., Cushings disease, Addison's disease, hypopituitarism, adrenal adenoma, diuretic abuse, cardiac failure. The purpose of today's experiment is to demonstrate the formation of the adrenal steroids by an extract of bovine adrenal glands and the inhibition of their formation by the drug Metyrapone. This drug is used in the clinic to interfere with adrenal steroid formation and thus test for the ability of the patient to secrete the pituitary hormone, adrenocorticotropic hormone (ACTH).

Principles

Most of the adrenal cortical hormones are present in relatively small amounts in the blood and are ring A-reduced dihydro- or tetrahydro derivatives of the major biologically active hormones or their intermediates. The biosynthesis of some of these hormones is outlined in Figure 10.1. The adrenal steroid hormones are classified as glucocorticoids, mineralocorticoids, and adrenal androgens.

Glucocorticoids are C_{21}-adrenal steriods that have an effect on carbohydrate metabolism, resulting in an increased blood-glucose level. *Cortisol* is the principal glucocorticoid secreted by the human adrenal cortex (15 mg/day).

Mineralocorticoids are C_{21}-adrenal steroids that have an effect on plasma electrolyte levels as a result of the kidneys' increased retention of Na^+ ions accompanied by their increased excretion of K^+ ions. *Aldosterone* (so named for its C_{18}-aldehyde group) is the principal mineralocorticoid secreted by the human adrenal cortex (0.1 mg/day).

Adrenal androgens are C_{19}-adrenal steroids that have a 17-ketone group and are weakly androgenic, i.e., they can be converted peripherally to the biologically active male steroid sex hormone, testosterone. The principal adrenal androgen secreted by the adrenal cortex (10 mg/day) by both males and females is *dehydroepiandrosterone*. An important feature of this substance is that it is secreted as a sulfate ester by the adrenal cortex and is present in circulating plasma in larger amounts than any other steroid hormone ($\sim$ 150 μg/dl) as dehydroepiandrosterone sulfate.

In humans, as in other animals, the adrenal cortex is stimulated by ACTH, a specific protein secreted by the pituitary gland. The administration of ACTH causes an increased rate of secretion of the adrenal cortical steroid hormones. Such a stimulation can be used clinically to evaluate the response of the normally functioning adrenal gland. The rate at which ACTH is secreted by the pituitary is kept in balance by the levels of cortisol circulating in the blood. The inhibitory effect of the plasma cortisol on pituitary ACTH secretion and the stimulating effect of ACTH on adrenal secretion of cortisol form the essential features of a negative feedback mechanism.

A releasing factor for ACTH, corticotropin releasing factor (CRF), is secreted by the hypothalamus. The CRF secretion rate is regulated by a combination of factors, including neurohumoral agents and plasma cortisol levels. Extracts of hypothalamic tissue containing CRF have been used to stimulate the pituitary to secrete ACTH, and the exact chemical structure of CRF has now been established. Because purified CRF is not widely available, it is not now possible to stimulate pituitary secretion of ACTH directly for clinical analysis. However, when CRF becomes commer-

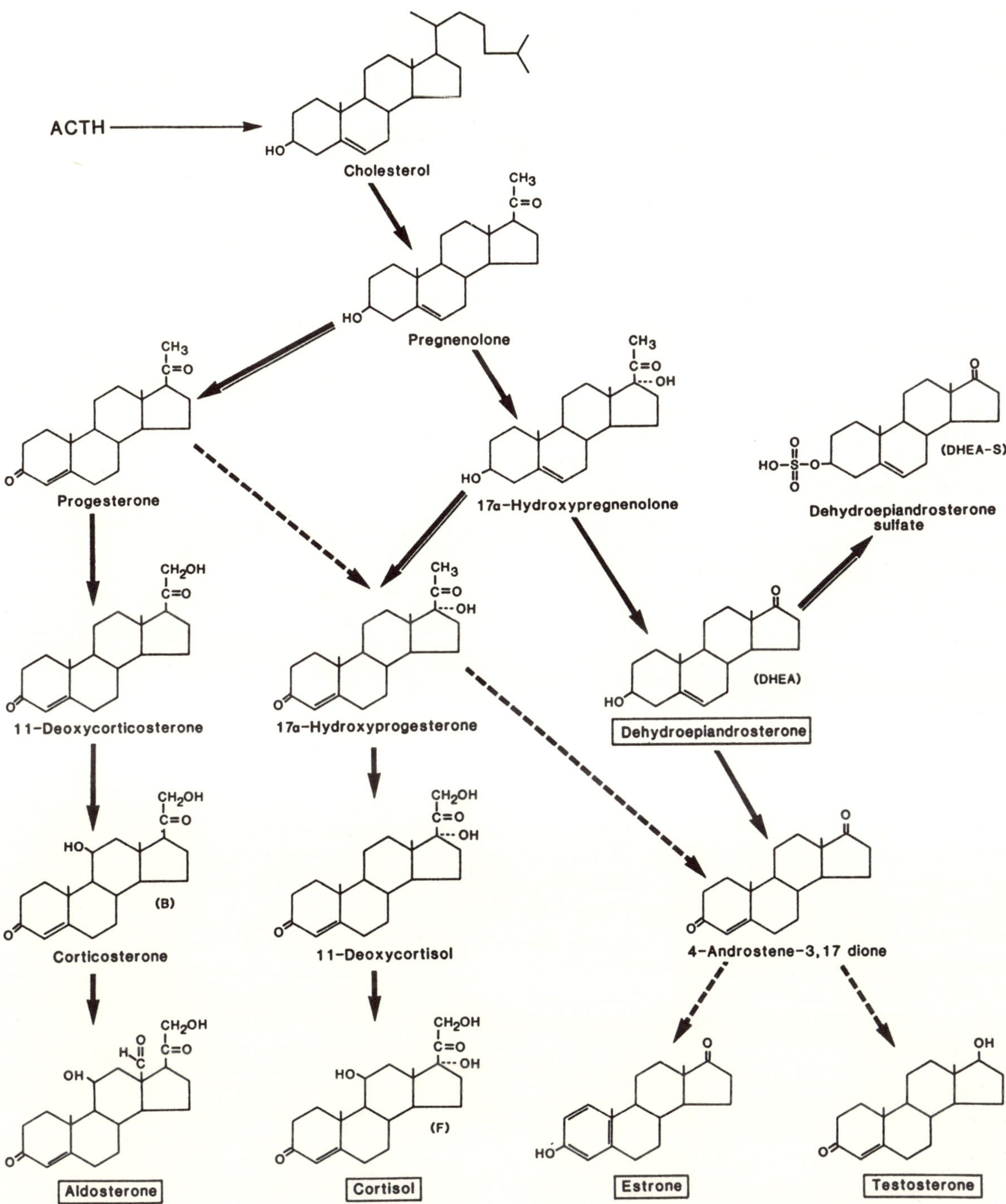

Figure 10.1. Major pathways in steroid hormone biosynthesis. Dotted lines indicate alternate pathways. Double lines indicate major branch points leading to final adrenal steroid hormones.

cially available, a direct test for ACTH secretion will be possible. At present the drug Metyrapone (SU−4885) is used in patients as an indirect means of estimating the ability of the pituitary to secrete ACTH (ACTH reserve). Metyrapone is one of several drugs that can be used clinically to interfere with the formation of steroid hormones by blocking the hydroxylation reactions involved in the synthesis of cortisol and aldosterone, the principal adrenal cortical hormones. The inhibition of hydroxylation reactions results not only in a decreased synthesis of cortisol, but also in the accumulation of those steroid components that are normally trace intermediates prior to the hydroxylation steps in the biosynthetic reactions. When Metyrapone is administered to a patient, the formation of cortisol is blocked. As a result, in individuals with a normal pituitary gland, the rate of ACTH secretion is increased; this stimulates adrenal secretion of steroid precursors and increases the size of the adrenal cortex. In individuals with an impaired ability to secrete ACTH, the secretion of adrenal steroid precursors is not increased.

Clinical Applications

Diseases of the Adrenal Cortex

Hypoactivity

The adrenal cortex is essential for life, but in the absence of adrenal cortical steroids, as a result of Addison's disease or following bilateral adrenalectomy, patients can be maintained quite adequately by the administration of just two substances, cortisol and aldosterone. At the present time a number of synthetically modified orally active steroids are commonly used clinically for replacement therapy as glucocorticoids or as mineralocorticoids.

Hyperactivity

Hyperactivity of the adrenal cortex can result in the abnormal state *Cushing's* syndrome, which is due to the excessive secretion of the glucocorticoid cortisol. A similar abnormal state (iatrogenic) can be induced with prolonged administration of cortisol or one of its synthetic analogs. When excessive secretion by the adrenal cortex is primarily confined to aldosterone, the abnormal state is referred to as *aldosteronism*. Increased androgenicity and virilism are observed with adrenal hyperactivity when excessive adrenal androgens are secreted. The *adrenogenital* syndrome (a congenital defect) refers to patients with excessive androgen stimulation with sequellae including abnormal genitalia.

Steroid Tests

Although the major adrenal steroid secretory products now are measured by very specific and sensitive radioimmunoassay procedures (RIA), several quantitative chemical tests are still being used to assay specific steroid products. Since the results of such assays will be part of the patient's clinical record, you should be familiar with names and specificities of these assay procedures. These assays are described in the Appendix to this chapter.

Procedures

The experiment outlined below illustrates the rationale for the use of Metyrapone and demonstrates the principal biosynthetic pathways for the formation of the adrenal cortical steroid hormones. In one flask a steroid hormone precursor [4-^{14}C]-progesterone is incubated with an enzyme preparation consisting of an acetone powder of bovine adrenal cortex. In a second flask Metyrapone (SU—4885) is added as well as [4-^{14}C]-progesterone. The effect of Metyrapone on progesterone metabolism is determined by chromatographic separation and quantitation of metabolic products formed in the absence and presence of the drug.

Materials

Two silica gel thin-layer chromatography strips (1.5 × 7.0 cm)

[4-^{14}C]-progesterone: (50 μCi/0.30 mg) in methanol (use .05 ml or .05 μCi)

NADPH:1 mg/ml in .02 M potassium phosphate buffer (use 0.2 ml)

Metyrapone (SU-4885): 0.1 mg/ml in potassium phosphate buffer (use 0.5 ml)

Enzyme: 10 mg/ml buffer, acetone powder of bovine adrenal cortex (use 1.0 ml)

Standard solution of a mixture (1.0 mg/ml methanol) of:

progesterone	(P)
11-deoxycorticosterone	(DOC)
corticosterone	(B).

Incubation

Label two 25 ml Erlenmeyer flasks 1 and 2. Add the following to the flasks in the order given.

Flask 1	0.05 ml[4-^{14}C]-progesterone
	0.5 ml potassium phosphate buffer
	1.0 ml enzyme solution
	0.2 ml NADPH
Flask 2	0.05 ml [4-^{14}C]-progesterone
	0.5 ml Metyrapone (SU-4885)

1.0 ml enzyme solution

0.2 ml NADPH

Place flasks in 37°C H_2O bath and incubate with shaking for 30 minutes.

Extraction

Remove flasks from incubator and add 0.02 ml mixture containing the three standard steroids (progesterone, DOC, and corticosterone). Add 6 ml methylene dichloride dispensed from a repipette. Swirl flasks gently for two minutes. Then pour into 12 ml thick-walled conical centrifuge tubes (labeled 1 and 2) and centrifuge at three-fourths speed for five minutes. Remove the top aqueous layer including residue between layers by aspiration with a water vacuum pump. The lower CH_2Cl_2 layer containing the steroid products is evaporated by air jet in a warm bath down to a small volume at the tip of the centrifuge tube. The sides of the tube are washed down with 0.10 ml CH_2Cl_2 and evaporated to dryness.

Chromatography

Add 0.02 ml acetone to the bottom of the centrifuge tube. Transfer residue one drop at a time to a point 1.5 cm from the bottom of a silica gel strip labeled 1, using the tip of a Pasteur pipette. Repeat transfer with another .02 ml acetone. Follow the same procedure using tube and strip labeled 2.

Allow silica gel strips to air dry. After samples have been applied and strips air-dried, dip the bottom edge of the strips in acetone and allow solvent to rise 2 cm to concentrate samples in a thin straight line. Place slides in covered glass developing container to which has been added 2.5 ml developing solvent (cyclohexane: ethylacetate 1:1). When solvent is three-fourths of the way up the strips, remove the strips and air dry. Observe U.V. absorbing zones in dark box and mark zones with pencil (see Figure 10.2). After chromatography the two strips look slightly different because the inhibitor (SU−4885) absorbs U.V. light, producing a dark zone (slide 2) between the B and DOC zones. For consistency this zone marker is included with the DOC band as shown; it does not interfere with the radioactive counting data.

The zones are cut and placed into liquid scintillation vials marked 1-P, 1-D, 1-B; 2-P, 2-D, 2-B, respectively, for the 1-series and 2-series, and 10 ml of counting fluid (PPO + POPOP in toluene) are added. The lists of counts and background counts are posted. Each student should obtain and record the results of incubations.

Selected References

Dorfman, R. I., and Ungar F. 1965. *Metabolism of Steroid Hormones*. New York, Academic Press.

Ungar, F. 1982. Biochemistry of Hormones I: Hormone Receptors; Steroid and Thyroid Hormones. In *Textbook of Biochemistry with Clinical Correlations,* ed. T. M. Devlin, pp. 714–755. New York, Wiley.

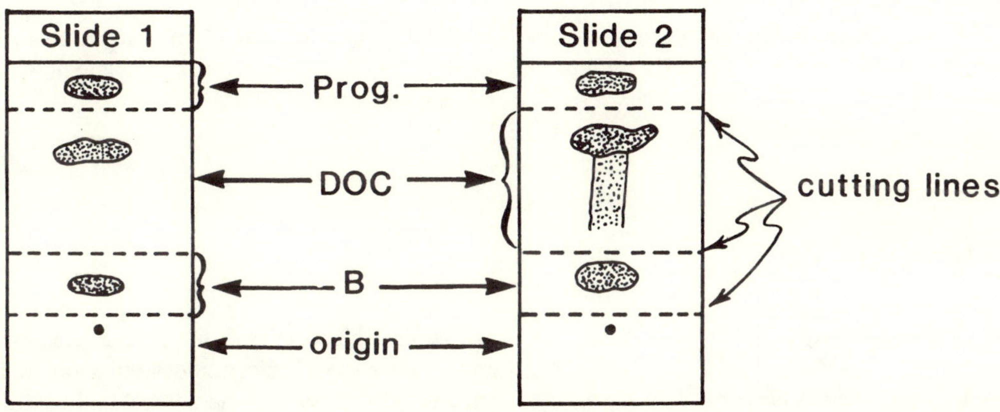

Figure 10.2. Thin layer chromatography of steroid hormone conversion products.

Problems

1. There are several forms of congenital adrenal hyperfunction, referred to as the adrenogenital syndrome. In this form of genetic disease an adrenal enzyme may be lacking, or deficient, which normally would hydroxylate the steroid at position 11, 21 or 17.

   a. What steroid(s) would be produced and secreted in largest amount if the enzyme abnormality involved a deficiency of (1) 11β-hydroxylation, (2) 21-hydroxylation, (3) 17-hydroxylation? (Refer to Figure 10.1.)

   b. List one or more of the physiological features which you would expect to be the most prominent as a result of the steroid blockage and the accumulation of these steroid intermediates.

2. Describe, in terms of the classical negative feedback system, what would occur in each case:

   a. Size and function of the adrenals two weeks after hypophysectomy.

   b. Size and function of the adrenals after daily administration of ACTH.

   c. Size and function of the adrenals after daily administration of cortisol (25 mg/day).

3. Refer to the metabolic map on p. 71 illustrating the steroidogenic pathways in the adrenal cortex.

   a. What steroid C_{21} compound is a precursor for all of the adrenal steroid products?

   b. The formation of C_{19} or androgenic steroid requires a hydroxylation at which carbon?

   c. The glucocorticoids (C_{21} steroids secreted by the adrenal cortex) contain hydroxyl groups at which two carbons?

   d. Cortisol contains a hydroxyl group at which additional carbon?

   e. What functional group at carbon 18 makes aldosterone a unique steroid?

4. ACTH and angiotensin both stimulate the human adrenal cortex to produce steroid hormones. What is the major steroid produced as a result of ACTH stimulation? What is the major steroid produced as a result of angiotensin stimulation?

5. Record the count rate and calculate the percent recovery of the compounds listed below. (Refer to Figure 10.2.)

Compound	*Count Rate (cpm)*	*% Recovered of Total*
1-P		
1-D		
1-B		
1-Total		
2-P		
2-D		
2-B		
2-Total		

6. What step in biogenesis of adrenal cortical hormones does Metyrapone block? Explain how the data support this conclusion.

Appendix

The 17-Ketosteroid Assay (Zimmermann Assay)

Any 17-ketone produced by the adrenal, ovary, or testis adds to the chromogenic value of this test. A purple color is produced by the 17-ketone in the presence of m-dinitrobenzene, 2.5 N KOH in absolute ethanol. The test measures metabolites of dehydroepiandrosterone, testosterone, estradiol, and 17α-hydroxysteroids. The test is not a good measure of testicular or ovarian secretory activity, since the bulk of 17-ketosteroids excreted in the urine are derived from the adrenal gland in both males and females. However, under specified conditions the assay is used to measure the activity of the adrenal and/or the testis. (See *Chem. Pharm. Bull*, 13:78, 1965.)

The Kober Assay

Estrogens contain a phenolic ring A, which is acidic and can be selectively removed from the neutral steroids by extraction with 1 Normal KOH. This fraction reacts with a 70% sulfuric/30% water reagent to produce a pink color (Kober chromogen), which measures individually (or a mixture of) estradiol, estrone, and estriol. The chromogen has fluorescence properties; and a more sensitive fluorescence test can be applied as well. The assay is used as a measure of ovarian activity in problems related to gynecology and obstetrics. (See *J. Endocrin.* 20:331, 1968)

Porter-Silber Assay (17α-Hydroxy Test, α-Ketol Test)

Steroids that contain a dihydroxy 20-ketone (α-ketol, 17α-hydroxy, 20-ketone, 21-hydroxy) side chain produce a yellow color in the presence of phenylhydrazine in concentrated sulfuric acid. Since the major steroid containing this particular C_{21}-side chain is cortisol or its metabolites, this test is relatively specific for adrenal secretion of cortisol. The steroids cortisone and 11-deoxycortisol also are chromogenic, but they normally are present in lesser amounts compared to cortisol. The compounds corticosterone and aldosterone, which lack the 17α-hydroxy group, are not chromogenic. The C_{21}-methyl steroids are not chromogenic. This assay is commonly used to measure the activity of the adrenal cortex. (See *J. Biol. Chem.* 185:201, 1950.)

The 17-Ketogenic Assay (Sodium Bismuthate Test)

Just as the 17-ketosteroids are derived from 17α-hydroxy-C_{21}-steroids during tissue metabolism, the chemical conversion of C_{21}-steroids to C_{19}-steroids (loss of carbons 20 and 21) proceeds more readily when a 17α-hydroxyl group is present. Sodium bismuthate is one of several oxidizing agents that cleave the C_{21}-side chain containing a 17α-hydroxy group to a 17-ketosteroid. The 17-ketosteroid test is described above. C_{21}-steroids with the C_{21}-side chains shown in Figure 10.3 can be converted to the 17-ketosteroids by using sodium bismuthate and other modifications (reduction for example) to increase the 17-ketogenic activity of the C_{21} side chain. Note that this test measures both cortisol and 17α-hydroxy-C_{21}-CH_3 steroids or their metabolites, and is used to measure the activity of the adrenal cortex. (See *Biochem. J.* 60:460, 1955.)

These four colorimetric assay procedures can be applied to both blood and urine after solvent extraction and some preliminary purification procedures. Most of the steroids excreted in urine, however, are reduced metabolites conjugated to glucuronic acid or sulfate groups. Therefore it is necessary to hydrolyze the conjugates after solvent extraction to free steroids before the colorimetric assays can be used. Two other assays for steroid hormones are available for qualitative and quantitative measurement.

1. The $\Delta^4 - 3$-ketone steroid hormones absorb in the ultraviolet region at 240 nm, whereas the $\Delta^5 - 3\beta$-hydroxysteroid precursors such as cholesterol, pregnenolone, and dehydroepiandrosterone do not absorb at 240 nm. The assay is used frequently to measure the concentration of pure steroids in aqueous solutions.

2. The 11β-hydroxysteroids, cortisol, and corticosterone, when dissolved in concentrated sulfuric acid-ethanol (65:35) have characteristic fluorescence spectra (excitation maximum, 465 nm; fluorescence maximum, 520 nm) which can be used on suitably purified steroid fractions. The 11-ketone forms, or ring A-reduced forms, do not fluoresce. This assay is used with plasma to measure the activity of the adrenal cortex.

The structures discussed in the above assays are shown in Figure 10.3.

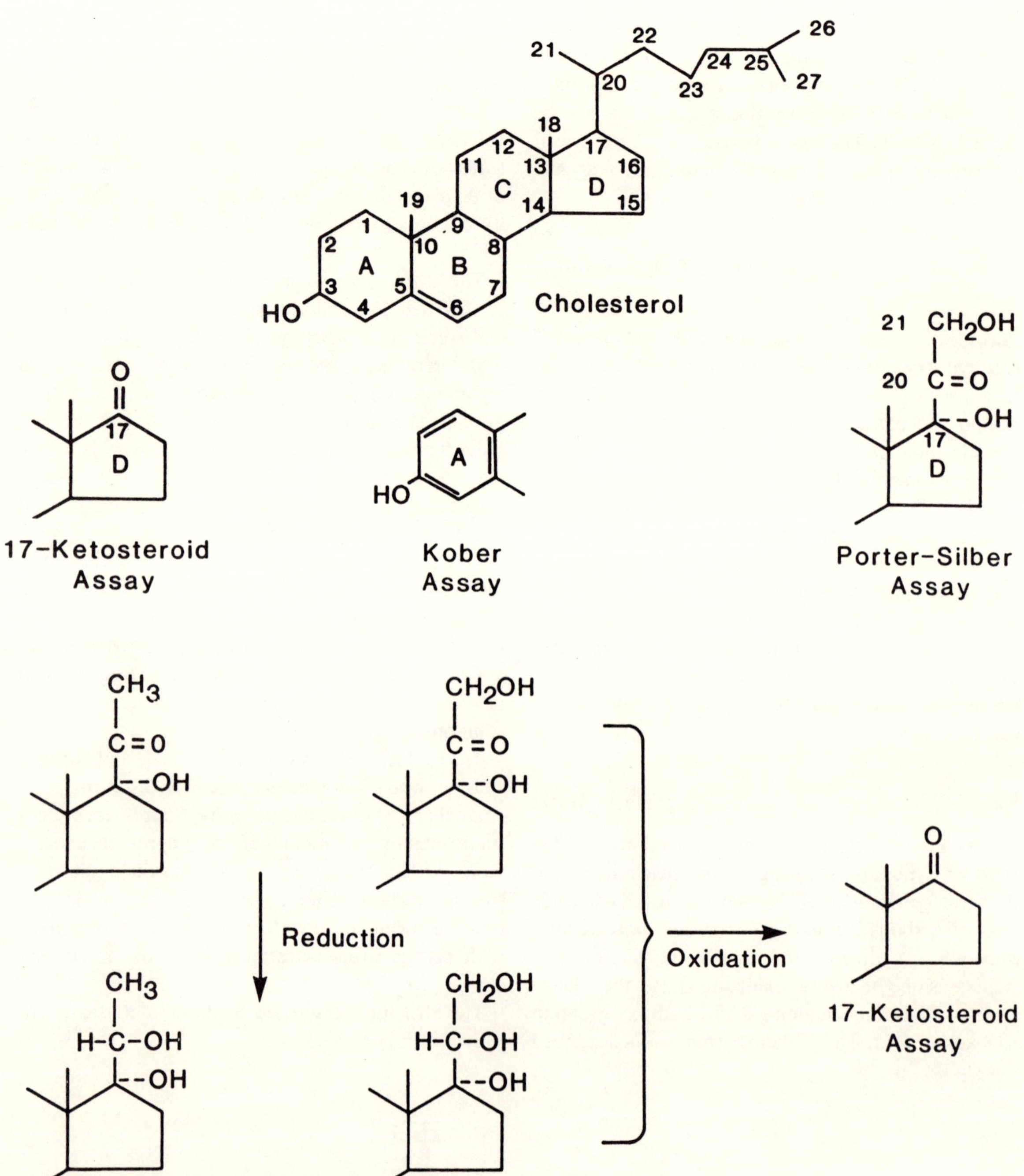

Figure 10.3. Colorimetric assay procedures for major groups of steroid hormones. The structure and numbering system for cholesterol is at the top. The four structures at the lower left are 17-ketogenic, i.e., they can be oxidized and assayed as 17-ketosteroids.

Immunoelectrophoresis of Serum Proteins 11

Maureen A. Scaglia and James F. Koerner

Only five protein fractions in human serum can be identified by the technique of simple electrophoresis. However, with the technique of immunoelectrophoresis, 15–20 proteins are identifiable. The serum proteins are separated first by electrophoresis in agar gel. Immunological differentiation of the separated proteins is achieved by adding specific antisera to a groove in the agar plate beside the separated protein fractions. Proteins and antisera diffuse through the agar and within 12 to 18 hours form precipitin bands. Immunoelectrophoresis is essential in the diagnosis of diseases of the immunoglobulin system.

Principles

The separation of three serum proteins by electrophoresis, followed by their diffusion into a gel containing antisera to normal human serum, is illustrated in Figure 11.1.

The position and shape of these precipitin bands is determined by the electrophoretic mobilities, rate of diffusion, and immunological specificity of each protein. With the exception of immunoglobulins (the γ globulins), each band is symmetrical. Each class of immunoglobulin is a highly complex mixture of proteins of different primary structures having discrete electrophoretic mobilities. Most available antisera detect only the major antigenic groups common to all members of a particular class, showing only a single band for each class. The size and intensity of a precipitin band is roughly proportional to the quantity of antigen. However, the amount of antigen cannot be quantitated by this method.

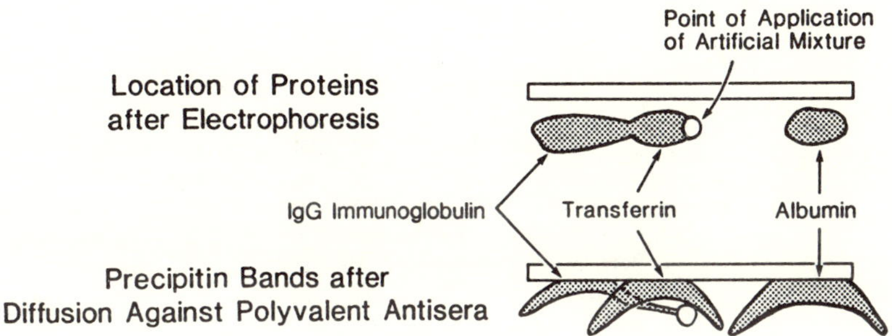

Figure 11.1. Separation of three proteins by immunoelectrophoresis.

The Process of Immunoprecipitation

The ability of immunoelectrophoresis to detect dozens of serum proteins is based on the selectivity and sensitivity of a technique called immunoprecipitation. If a pure protein, called an antigen, is injected into an animal such as a rabbit or guinea pig, its immune system responds by producing specific antibodies to the protein. An appropriate mixture of antigen and antibody-containing serum from the animal forms a precipitate. However, if there is an excess of either antigen or antibody, little precipitate forms. This is illustrated by the graph in Figure 11.2, which shows the effect of adding increasing amounts of antigen to a constant amount of antibody. At the molecular level, only when antigen and antibody molecules are present in approximately equivalent amounts do the two kinds of molecules interact to form a high-molecular-weight, insoluble aggregate (Fig. 11.2).

After the electrophoresis of a protein (antigen), antibody is added to a slot in the electrophoresis slide, which runs parallel to the direction of electrophoresis. During the next few hours the antigen and antibody

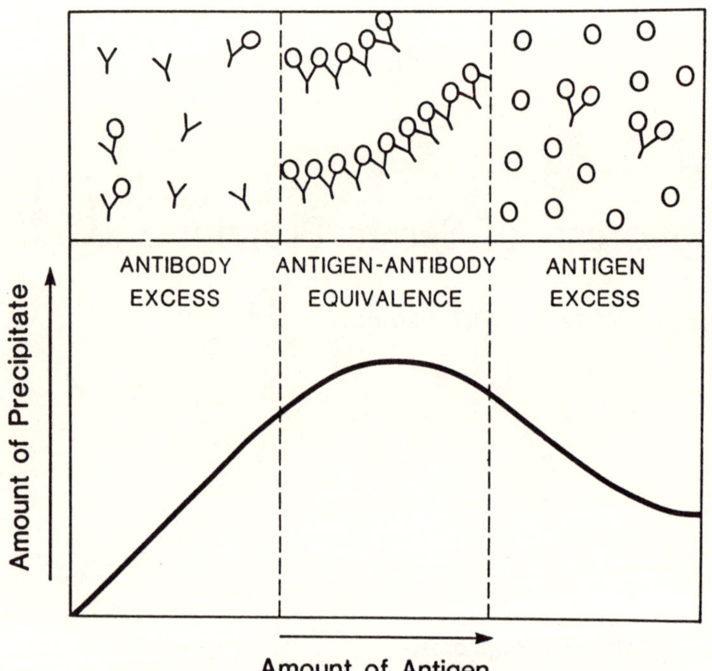

Figure 11.2. Bottom: Graph of relationship between amount of antibody added and immunoprecipitate formed. Top: Molecular interactions under conditions of antibody excess, equivalence and antigen excess. O = antigen molecule, Y = antibody molecule.

molecules diffuse through the agar. As shown in Figure 11.3, a single protein-antigen spot diffuses radially in all directions, whereas, the antibody molecules in the slot appear to diffuse as an advancing linear front. These macroscopic effects are the result of many individual molecules of the proteins diffusing in random paths. When the merging antigen and antibody attain equivalence, the insoluble complex is deposited in the agar. As this process continues, an arc of precipitate forms between the antigen and antibody wells (Figure 11.3). If antigen is in excess, a thick arc of precipitate forms near the antibody well and smears into the well. The albumin precipitate obtained by immunoelectrophoresis of serum has this appearance. If antibody is in excess, the arc forms near the antigen spot and is small and diffuse because of poor cross-linking between antigen and antibody molecules (Figure 11.4).

Serum protein immunoelectrophoresis patterns are usually developed with antibodies prepared by injecting an animal with whole human serum. Serum from such animals contain antibodies against most of the

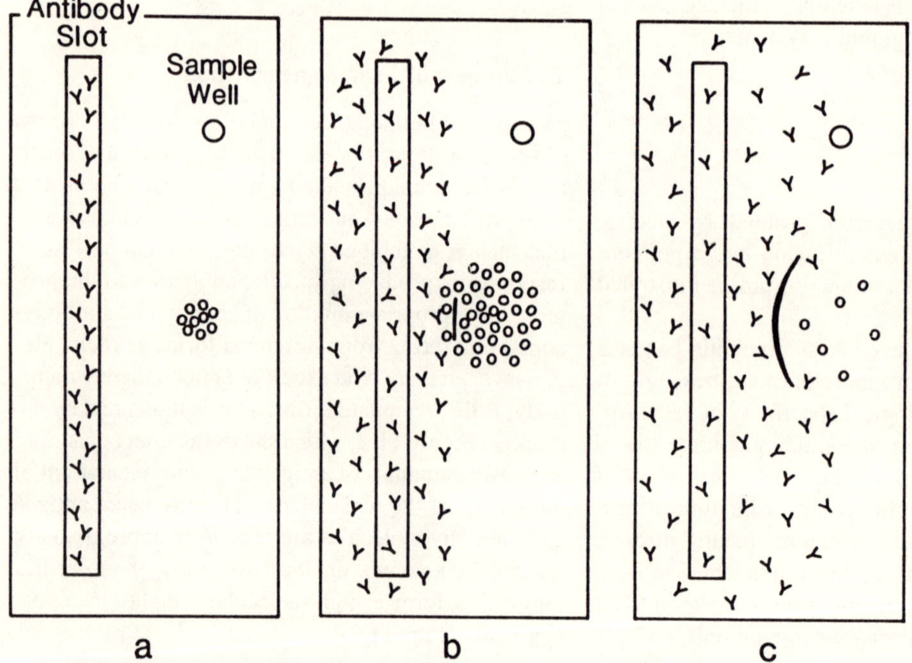

Figure 11.3. The formation of a precipitin line. a) Immediately after electrophoresis. Antigen spot (o) and antibody (Y) added to the spot have not diffused extensively. b) Short time, precipitin arc beginning to form. c) Still later-precipitin arc formed.

human serum proteins. The resulting immunoelectrophoresis pattern is a complex pattern of arcs, each representing a specific protein. Antisera for specific proteins or classes of proteins can also be used, giving a simpler pattern that emphasizes the desired proteins.

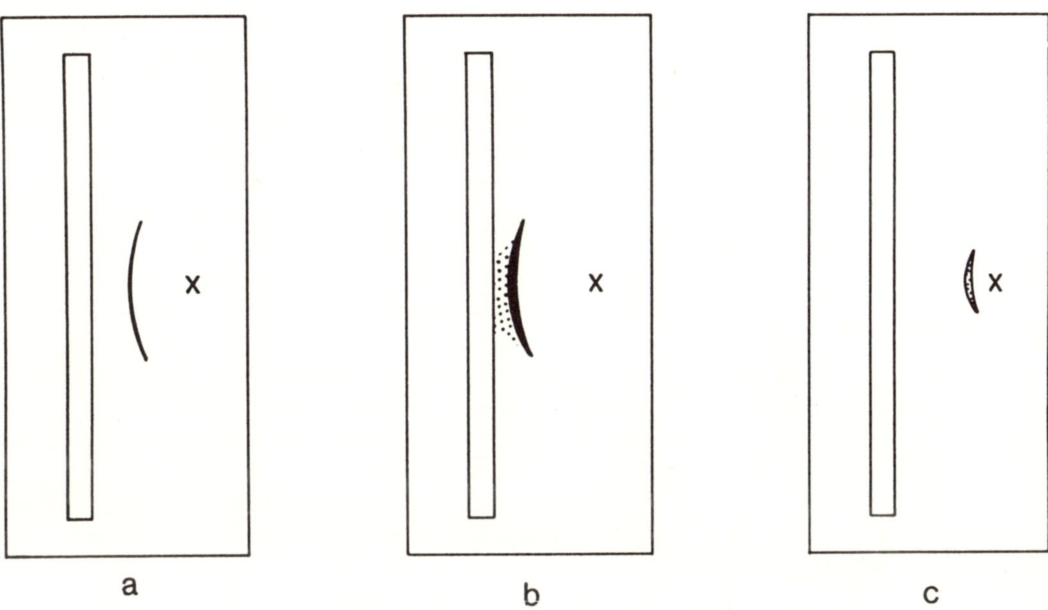

Figure 11.4. The influence of antigen and antibody concentrations on precipitin arc formation. a) Standard antibody-antigen ratio showing a clear, narrow arc of precipitation. X = position to which antigen migrated during electrophoresis. b) Excess antigen showing heavy precipitin arc near antibody slot. c) Excess antibody showing weak precipitin arc near antigen spot.

Properties of Immunoglobins

The structures of the immunoglobins are discussed in your textbook. Some properties of the immunoglobins are listed in the table at the end of this chapter.

Three of the immunoglobin classes, IgG, IgA, and IgM, are readily detected by immunoelectrophoresis. Each of these classes gives an elongated line of precipitation, unlike the arc shown by a single pure protein. This is because IgG, for example, is not a single protein but a mixture of thousands of closely related immunoglobulin proteins. All IgG protein molecules have similar constant regions in their amino-acid sequence. It is these constant regions that act as antigenic determinants. When an IgG mixture is injected into an animal, antibodies are developed which will react with the constant regions of all molecules of the IgG class. The IgG molecules also have regions in which the amino acid sequence differs from other proteins of the class. These variable regions give the immunoglobulins their differing specificities as disease-fighting antibodies. They also give the molecules different electrophoretic mobilities. Because each individual immunoglobulin is present in a minute amount, the variable regions do not elicit enough antibodies in animals to produce a serum that forms an immunoprecipitate with the antigen.

Each IgG molecule of a certain exact structure is the product of a clone of cells. This clone arose from a single cell that proliferated in response to stimulus by an antigen. Thousands of these clones produce the IgG fraction in human serum. Immunoelectrophoresis of this fraction gives the elongated band of immunoprecipitate shown in Figure 11.5, where the mobilities of 10 different IgG proteins are diagrammed.

If a single IgG producing cell proliferates wildly, it can produce a disease called multiple myeloma, a cancer of the immunoglobulin system. The serum from such a patient contains a massive amount of a single kind of IgG molecule. The band of IgG precipitate resulting from such serum is also illustrated in Figure 11.5. In many myeloma patients the cancerous cells crowd out other immunoglobulin-producing cells so the other immunoglobulin bands are reduced in intensity or missing entirely.

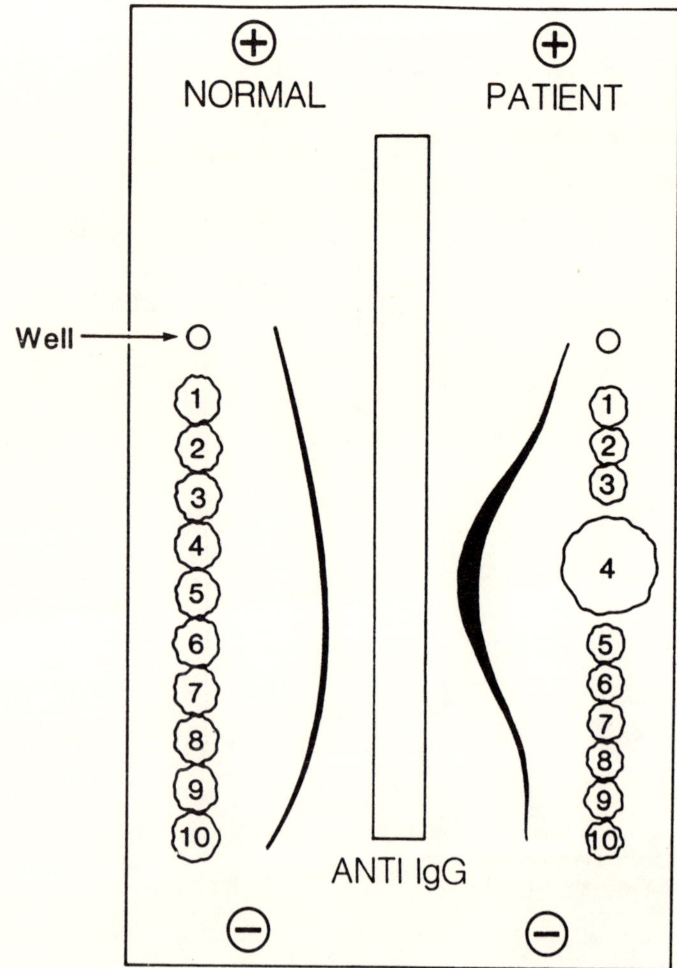

Figure 11.5. Immunoelectrophoresis of serum from a normal subject and a subject with myeloma who is overproducing a specific IgG protein. The migration pattern of ten specific IgG types are shown by ○

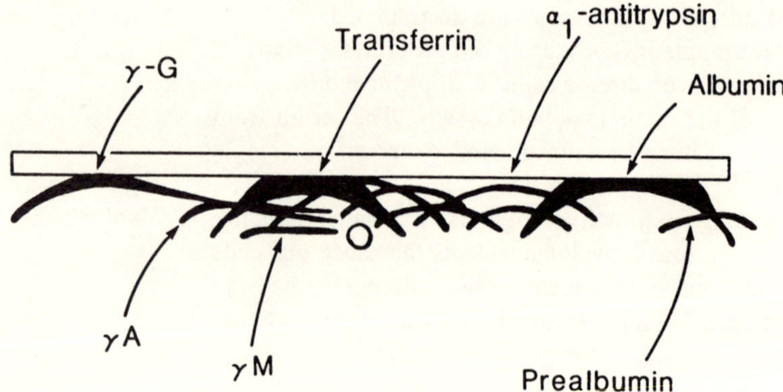

Figure 11.6. The pattern obtained after immunoelectrophoresis of normal serum (only major bands are shown).

Immunoelectrophoresis of Human Serum

Immunoelectrophoresis of the serum should always be done if any abnormal patterns are observed after *serum* electrophoresis in the α-, β-, or γ-globulin regions. A pattern obtained after the immunoelectrophoresis of normal serum is shown in Figure 11.6. The labeled precipitin bands are identified by their characteristic position (electrophoretic mobility) and shape (symmetry or asymmetry). The other numerous bands can only be identified by using specific antisera or a specific staining reaction. It is advisable to run a normal control with each antiserum trough used. Any change in precipitin size or shape due to test conditions can thus be detected. Often two concentrations of antihuman antiserum are used because the presence and amount of precipitate are dependent on antibody and antigen concentration. A band may appear or be better defined with one concentration than with the other.

Normal patterns seen with specific antisera to the immunoglobulins are shown in Figure 11.7. (The anti-Ig antiserum contains antibodies against all the heavy chains and the light chains.) Antiserum to human serum may be preferred to immunoglobulin (anti-Ig) antiserum in clinical use. Note that with good light chain antisera the normal kappa: lambda light chain ratio of 2:1 can be seen. There is not enough IgD or IgE present in normal serum to react with the antisera against them.

Immunoelectrophoresis can also be done on other body fluids. It is especially important to do a urine electrophoresis and immunoelectrophoresis if the possibility of a light chain gammopathy exists. The abnormal light chain may be present only in the urine and not in the serum. (Free light chains are also known as Bence Jones proteins.)

Clinical Applications

Changes in Immunoglobulin Levels as Detected by Immunoelectrophoresis

Low Levels of Immunoglobulin

If all classes of immunoglobulins in the patient are decreased, the *serum* electrophoresis pattern shows a decrease in the γ region. With *immunoelectrophoresis* the precipitin bands are smaller and/or shorter in comparison with a normal control. Although smaller, the reaction with the light chain antisera still reflects the normal 2:1 ratio of kappa to lambda light chain. In some congenital and acquired abnormalities, concentrations of one or two of the immunoglobulin classes

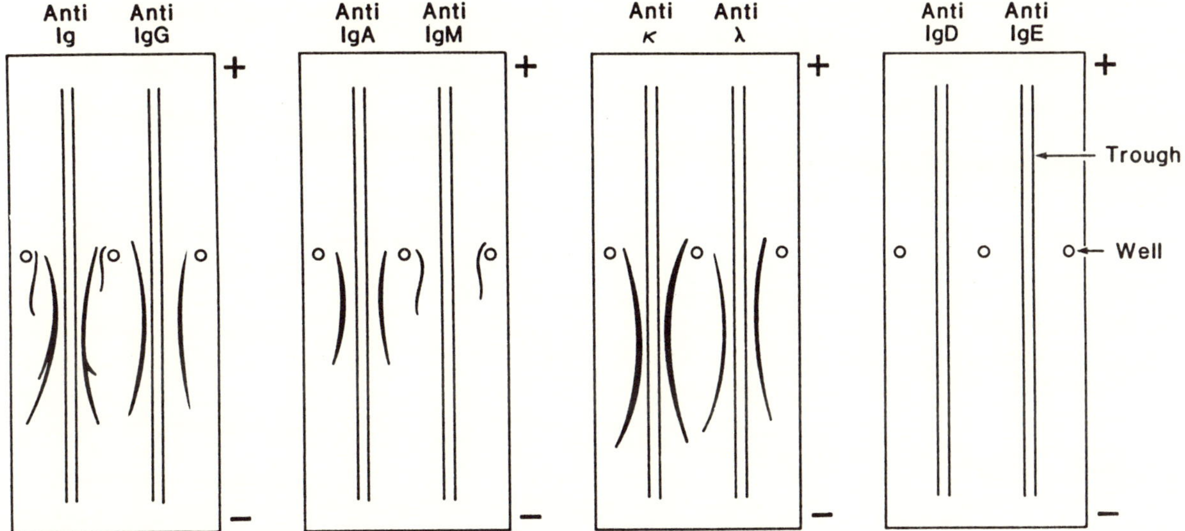

Figure 11.7. Immunoelectrophoresis of normal serum using anti-immunoglobin antisera

are decreased or absent. Thus the reaction with monospecific antisera for the affected class is decreased or not seen. Some conditions causing decreased levels of immunoglobulins include kidney disease, malnutrition, some cancers, and light chain (Bence Jones) monoclonal gammopathy.

Agammaglobulinemia

If no or very few immunoglobulins are being produced by the body, no precipitin reaction occurs with any of the heavy or light chain antisera. On *serum* electrophoresis, the γ region is very depressed or absent. Conditions giving rise to this state may be congenital (e.g., Bruton's agammaglobulinemia, see Figure 11.8), or acquired (e.g., certain cancers).

High Levels of Immunoglobulins

Polyclonal Gammopathy (many clones producing an excess of many different immunoglobulins). This condition is usually first detected on a serum electrophoresis showing a broad diffuse increase in the γ region. With immunoelectrophoresis, thicker and/or longer precipitin bands are seen with antisera against one or all of the heavy chains and with *both* light chain antisera. The normal kappa to lambda ratio is maintained. Collagen disorders, liver disease, chronic infection, and autoimmune diseases often lead to increased levels. Polyclonal free light chains may be seen in the urine, especially with lupus erythematosis and rheumatoid arthritis. The immunoelectrophoretic patttern seen in one type of polyclonal gammopathy is shown in Figure 11.9.

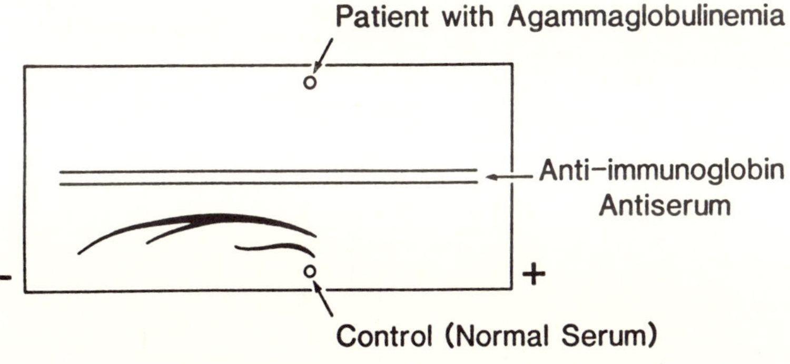

Figure 11.8. Immunoelectrophoresis of serum from patient with agammaglobulinemia

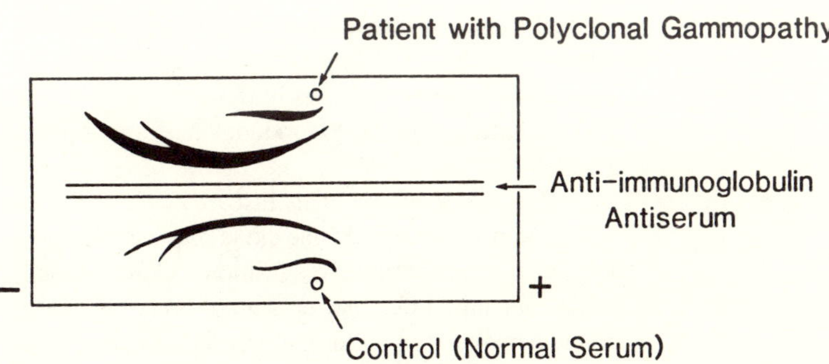

Figure 11.9. Immunoelectrophoresis of serum from patient with polyclonal gammopathy

Monoclonal Gammopathy (one clone of cells producing large amounts of one immunoglobulin, or a part of one). On serum electrophoresis a sharp increase or spike (M-peak) in one area of the β or γ region may be indicative of this abnormality. An immunoelectrophoresis must be done to classify this M-peak. The result will be an abnormal amount of precipitate and/or curve in one area of an arc in comparison to a normal control. Identification involves seeing the abnormal reaction with one of the heavy chain antisera and/or one of the light chain antiserum. In rare cases only a heavy chain is involved. Often the concentrations of the other immunoglobulins are decreased. (See Figure 11.10. Notice that the normal ratio of kappa to lambda light chain is altered. The report is: monoclonal IgA of lambda light chain type [IgA(λ)]; decreased IgG; no IgM seen.)

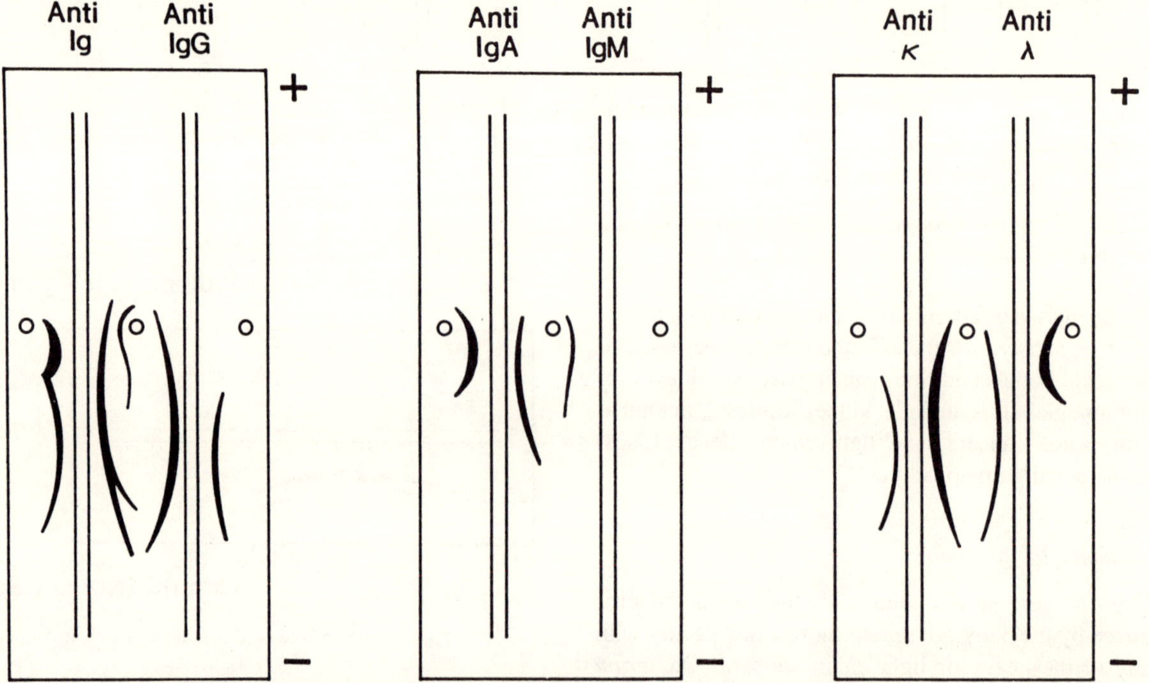

Figure 11.10. Immunoelectrophoresis of serum from a patient with a monoclonal gammapathy. Center well contains control (normal serum). Outer wells contain serum from patient with monoclonal gammapathy.

Conditions giving rise to a monoclonal gammopathy include:

1. *Multiple Myeloma*. This disease is the result of unlimited proliferation of one clone of plasma cells in a cancerous and invasive fashion. Consequences include anemia, infection, erosion of bone, and renal tubule damage. The abnormal protein may be a complete immunoglobulin—one heavy chain (rarely μ) and one light chain. The same light chain may be found in the urine (Bence Jones protein). Examples include IgD (λ), IgG (κ), IgE (λ). In other cases the abnormal protein is an incomplete immunoglobulin made up of just one of the heavy chains or just one of the light chains (free κ or free λ). In the latter case the free light chain is found in the urine but not always in the serum. Monoclonal gammopathies of the IgD or IgE class are rare, as are those producing just a heavy chain.

2. *Waldenstrom's Macrogammaglobulinemia*. This name is given to a cancerous disease state in which the monoclonal immunoglobulin is almost always an IgM (κ) or an IgM (λ). The cellular picture is usually predominately lymphocytic, although small plasma cells and intermediates may be seen. Symptoms can include hyperviscosity syndrome (retinopathy, neurological symptoms, etc.), anemia, hemorrhagic diathesis, weakness, and adenopathy. Since the IgM macroglobulin may possess cryoglobulin properties, intolerances to cold can also be manifested.

3. *Benign or Idiopathic*. A person with a monoclonal protein may be disease-free. This condition can be presymptomatic and should be monitored by serum and immunoelectrophoresis over a period of time.

4. *Other*. Other cancers or diseases (e.g., Lupus erythematosis, lymphoid malignancies) sometimes result in a monoclonal gammapathy.

Procedures

Immunoelectrophoresis of Plasma

Serum is usually used in the clinical laboratory. Plasma is used in this procedure to illustrate the presence of fibrinogen in plasma.

1. Racks of six agar plates are ready to be used by three pairs of students. The slides were first coated with an impregnation agar (to improve the adhesion of the agar to the glass slide) and then with 1.5% agar. Wells and troughs were punched in the film of agar. The wells are to be removed by suction, using a Pasteur pipette attached to a vacuum line.

2. One drop of 0.1% bromphenol blue is added to the plasma samples and the plasma is gently mixed. A Drummond pipette, set at 3 μl, is used to add the plasma to the wells. The formation of air bubbles in the wells should be avoided.

3. After all wells are filled, the frame is placed across the bridge of an electrophoresis chamber that has been filled with borate-phosphate buffer. Wicks that have been immersed in buffer are placed on the ends of the frame and into the buffer.

4. Electrophoresis is at 400 volts for about one hour, during which time the bromphenol blue spot should have migrated about 2 cm.

5. The power is turned off and the frames are removed from the chamber. The agar gel is lifted from the trough with a gel cutter and a tweezer.

6. The trough is filled with goat antiserum to normal human plasma with the aid of a 50μl pipette.

7. The frame carriers are placed in a plastic box containing a wet sponge. The presence of water prevents the dehydration of the agar gels. The plasma proteins separated by electrophoresis and the antiserum are allowed to diffuse toward one another in the agar gel for about 18 hours.

8. The laboratory staff immerses the gels in 1% saline solution to remove any protein present that has not formed an antigen-antibody complex. The gels are stained by the laboratory staff as follows:

   a. The gels are immersed in deionized water for one hour.

   b. The gels are covered with a damp filter paper, and the water in the gels is removed by evaporation with a stream of air from a fan.

   c. The gels are immersed in a solution of a protein dye, amido black, for 10 minutes.

   d. The dye that has not combined with the protein is removed from the gels by immersion in a mixture of acetic acid and methanol.

   e. The slides are removed from the frame and allowed to dry.

Selected References

Grant, G. H. and Kachmar, J. F. 1976. The Proteins of Body Fluids. In *Fundamentals of Clinical Chemistry*, ed. N. W. Tietz, 2nd ed., pp. 298–376. Philadelphia, Saunders.

Maldonado, J. E., Kyle, R. A., McDuffie, F. C., and Linman, J. W. 1973. Pathophysiology of the Monoclonal Gammopathies. *Postgraduate Medicine* 53-7:102–106; 54-1:139–145.

Markowitz, H., and Jiang, Nai-Siang. 1976. Immunochemical Principals. In *Fundamentals of Clinical Chemistry*, ed. N. W. Tietz, 2nd ed., pp. 274–297. Philadelphia, Saunders.

Penn, G. M., Batya, J., and Carl, L. Laboratory Evaluation of Immunoglobulins. *Laboratory Communications*. Vol. 1. Chaska, Minn., Kallestad Laboratories, Inc.

Riccardo, M. J. 1979. The Immunoglobulins: Biology and Structure. In *Todd-Sanford-Davidsohn: Clinical Diagnosis and Management by Laboratory Methods*, ed. J. B. Henry, 16th ed., Vol. 2, pp. 1214–1244. Philadelphia, Saunders.

Ritchie, R. F. 1979. Specific Proteins. In *Todd-Sanford-Davidsohn: Clinical Diagnosis and Management by Laboratory Methods*, ed. J. B. Henry, 16th ed., Vol. 1, pp. 228–258. Philadelphia, Saunders.

Tomar, R. H. 1979. Gammopathies, Hypersensitivity, Immunologic Deficiency. In *Todd-Sanford-Davidsohn: Clinical Diagnosis and Management by Laboratory Methods*, ed. J. B. Henry, 16 ed., Vol 2, pp. 1381–1408. Philadelphia, Saunders.

Problems

1. A mixture of five proteins is subjected to electrophoresis in agar gel at pH 8 and then allowed to diffuse against an antiserum to the five proteins. The immunoelectrophoresis pattern for the protein mixture is given in Figure 11.11. The molecular weights and isoelectric points of the five proteins are:

Protein	*Molecular weight*	*Isoelectric point*
A	25,000	5.5
B	100,000	4.5
C	250,000	6.5
D	150,000	8.5
E	50,000	6.5

Table 11.1. Properties of Immunoglobulins

	IgG	IgA	IgM	IgD	IgE
Heavy chain	γ	α	μ	δ	ϵ
Light chain	κ,λ	κ,λ	κ,λ	κ,λ	κ,λ
Molecular formula	$\gamma_2\kappa_2$ $\gamma_2\lambda_2$	$(\alpha_2\kappa_2)_n$ $(\alpha_2\lambda_2)_n$ n = 1,2,3	$(\mu_2\kappa_2)_5$ $(\mu_2\lambda_2)_5$	$\delta_2\kappa_2$ $\delta_x\gamma_2$	$\epsilon_2\kappa_2$ $\epsilon_2\lambda_2$
Molecular weight	160,000	170,000 (monomer form)	950,000 (pentamer form)	180,000	200,000
Designation	IgG(κ) IgG(λ)	IgA(κ) IgA(λ)	IgM(κ) IgM(λ)	IgD(κ) IgD(λ)	IgE(κ) IgE(λ)
Adult concentration (mg/dl)[a]	800–1600	50–400	40–250	0.5–3.0	0.01–0.04
Sedimentation	7S	7S, 9S, 11S, 13S	19S	7S	7–8S
Antibody functions and remarks	Comprise 80% of total antibody activity Antibacterial, antiviral, antitoxic, antinuclear activities Incomplete Rh antibodies Binds complement Only antibody capable of crossing placental barrier, protects newborns $T_{1/2}$ is 25 days, but catabolic rate is dependent on concentration	Principle antibody in organ systems lined with mucosal epithelium existing as a dimer of IgA linked by a "secretory piece" "Secretory-IgA" found in saliva, tears, colostrum, and mucous secretions of G.I., respiratory, and genital-urinary tracts Antimicrobial, does not bind complement IgA present in serum lacks "secretory piece" $T_{1/2}$ is 6 days, independent of concentration	First produced in response to antigenic stimulus Most primitive in terms of phlogeny and ontogeny Fixes complement the best Antibacterial (esp. gram-neg.), antiviral, antiparasitic (blood) Natural antibodies (ABO isoagglutinins autoimmune) High levels in cord blood of newborns suggestive of intrauterine infections $T_{1/2}$ is 5 days, independent of concentration	Unknown $T_{1/2}$ is 3 days	Increased in parasitic infections Involved in allergic immunity (reagenic, skin-sensitizing) $T_{1/2}$ is 2 to 3 days

[a]Normal values can vary with method used, sex, race, geographic location, and especially age. (In normal newborns only maternal IgG is present.) Production begins shortly after birth and rises to near-normal adult levels by 5 to 10 years of age.

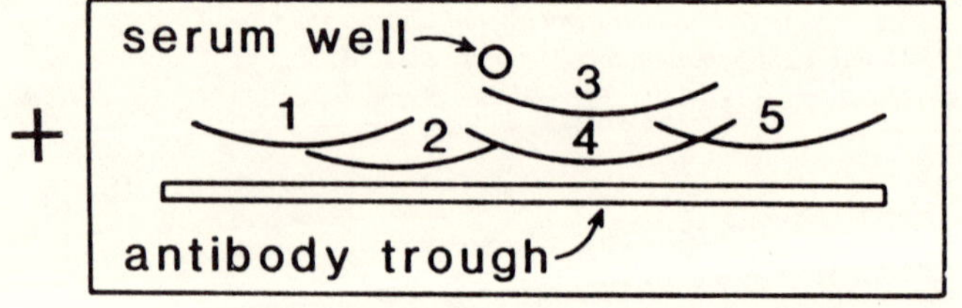

Figure 11.11. Immunoelectrophoresis of five proteins

List the protein positions according to the numbered position lines.

2. In the immunoelectrophoresis of human serum, the position of the IgM, IgA, and IgG precipitin lines relative to the sample well indicate that these proteins migrate toward the cathode. Explain this migration and the relative positions of IgM, IgA, and IgG.

3. Draw the pattern obtained with the immunoelectrophoresis of your plasma.

4. Give at least two reasons why some serum proteins are not visualized by immunoelectrophoresis.

5. Given these data:

Serum electrophoresis: normal.

Urine electrophoresis: abnormal peak in the gamma region, urine concentrated 10x.

Serum immunoelectrophoresis: normal IgA and IgG, no IgM seen.

Urine immunoelectrophoresis: urine concentrated 10×, the pattern seen is illustrated in Figure 11.12. What is the immunoelectrophoresis report? What is another name for the abnormal protein? Why is the abnormal protein's precipitin arc so close to the antiserum trough? What is happening in the patient's body, and what disease might this suggest?

6. The following data are obtained:

Serum immunoelectrophoresis is performed in the normal fashion.

The results with the antiserum to IgM and light chains are illustrated in Figure 11.13.

All other serum proteins are present in normal amounts except for IgA and IgG, which are present in low amounts.

The patient's serum is then treated with β-mercaptoethanol (MCE). MCE reduces the disulfide bonds in the IgM molecule, allowing it to break into its subunits, which are similar to IgG.

The result is given in Figure 11.14.

What is the immunoelectrophoresis report? What disease state is possible with these data? Explain the difference in reactions on the two sets of slides shown.

7. In the clinical laboratory, a normal control is run on each slide so that the control reacts with the same application of antiserum as the patient's serum. Why is this necessary?

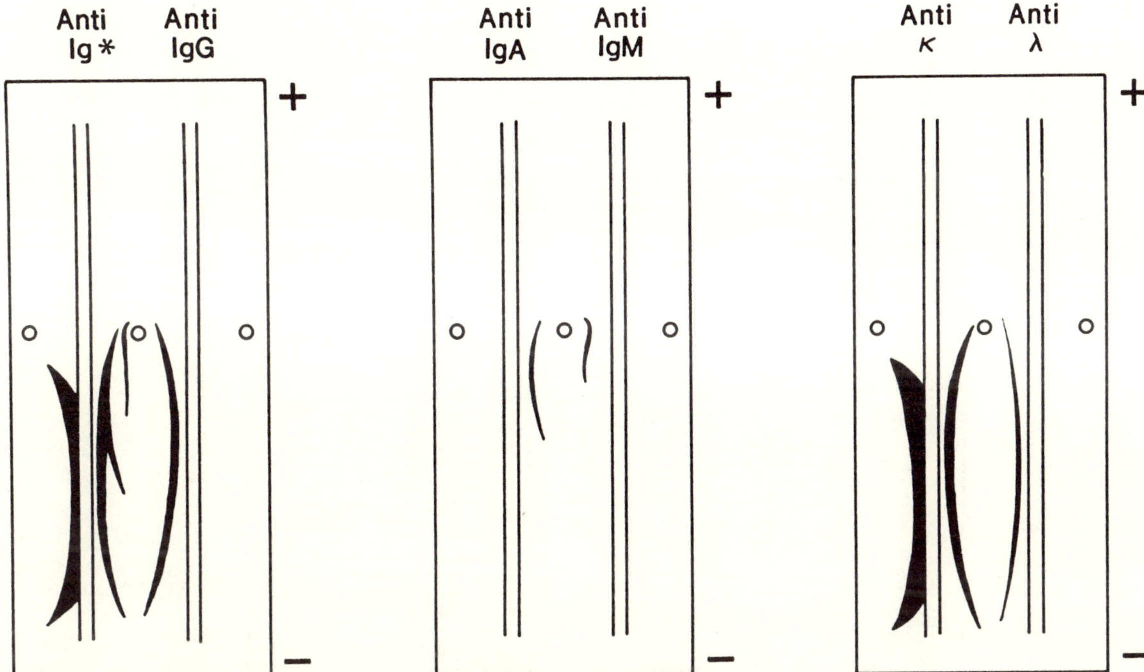

Figure 11.12. Immunoelectrophoresis of a patient's urine. Wells—Normal serum control in center well. Patient's urine in outer wells. *Antisera against all heavy chains and all light chains.

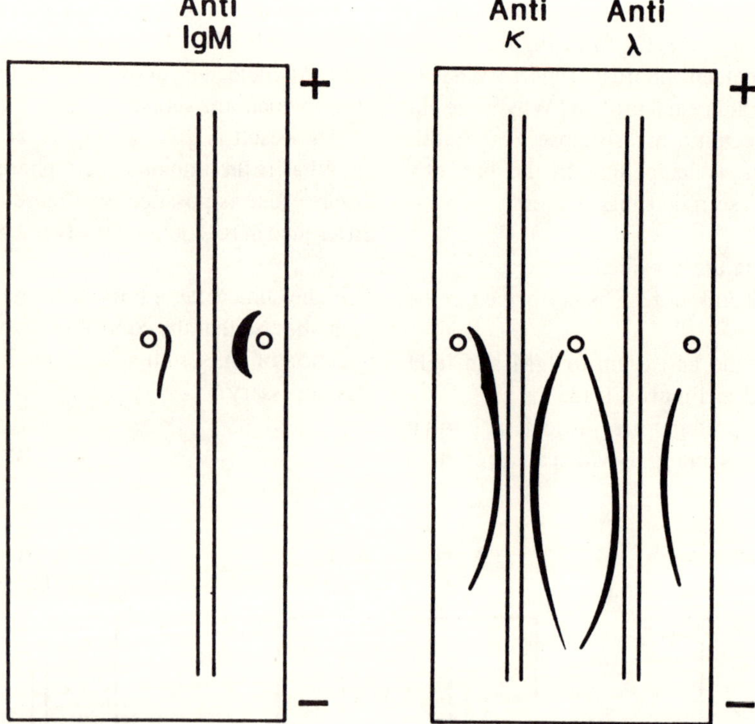

Figure 11.13. Immunoelectrophoresis of an abnormal serum. Wells—normal serum control in center, patient's serum in outer wells.

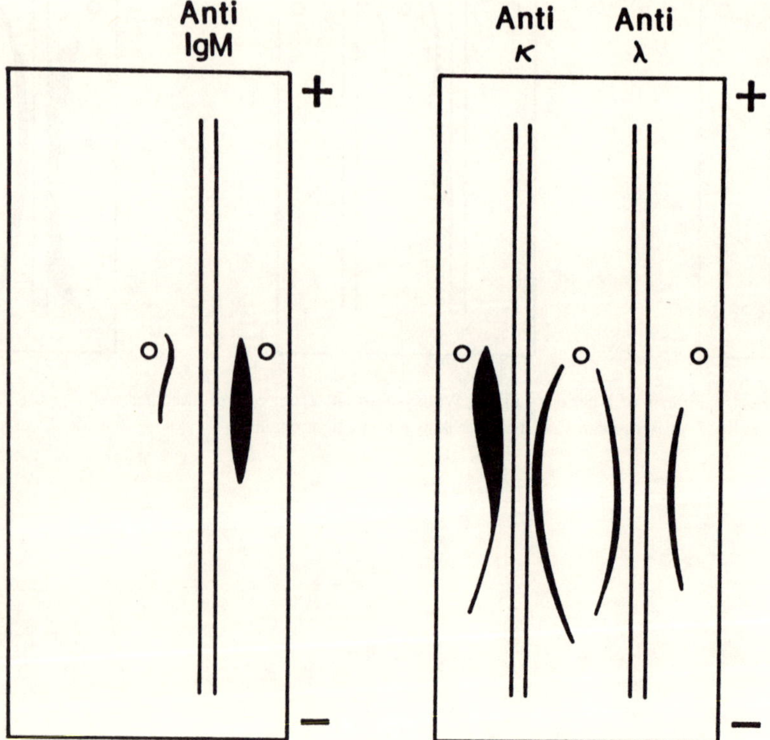

Figure 11.14. Immunoelectrophoresis of the serum seen in Figure 11.13 after treatment with MCE. Wells—normal serum control in center, patient's serum in outer wells.

Radioimmunoassay of Thyroxine 12

Frank Ungar and John F. Van Pilsum

Many compounds such as hormones, vitamins, and drugs are present in blood and tissues at such low concentrations that their quantification by chemical means, visible or ultraviolet spectroscopy, or even by biological assays is difficult. The levels of these compounds in tissues and biological fluids are of great diagnostic value, and the concentrations of a large number of these compounds can be ascertained *only* by radioimmunoassays (RIA) in which levels as low as 10^{-12}g (picogram) to 10^{-15}g (femtogram) may be detected. Some examples of substances that are measured by radioimmunoassays are:

- pituitary hormones
  - adrenocorticotropin
  - follicle stimulating hormone
  - luteinizing hormone
  - growth hormone
  - thyrotropin
- steroid hormones
  - cortisol
  - estradiol
  - progesterone
  - testosterone
- other hormones
  - gastrin
  - thyroxine
  - triiodothyronine
  - vitamin D
- hematologic compounds
  - erythropoietin
  - ferritin
- other compounds
  - many autoantibodies
  - myoglobin
  - several tumor markers
  - viral particles

Principles

Any protein, whether it is in the blood, within a membrane, in a cell such as a transport protein, an enzyme, or a drug or hormone receptor, can serve as a reagent for radioimmunoassays if it will bind the ligand molecule specifically. Moreover, the binding agent need not be a protein; it can be any material that can specifically bind a radioactive ligand, which is then displaced in the presence of the nonradioactive form of this ligand. An antibody to a pure antigen is the best example of a system in which a protein binds to a specific ligand. The radioimmunoassay of thyroxine performed today is an example of such a procedure. Note that the antibody is being used simply as a chemical reagent to bind thyroxine.

Foreign proteins injected into a rabbit, goat, or guinea pig generate antibodies which complex and inactivate the foreign protein. Large peptide hormones (such as growth hormone) and luteinizing hormone are antigenic and induce the formation of antibodies. Small peptide hormones (below 5,000 M.W.) are poor antigens. However, if the peptide is covalently linked to a larger protein molecule, the complex will be antigenic and will generate antibodies specific for the peptide as well as for the linked protein.

Nonpeptide hormones, steroids, thyroxine, drugs, and organic compounds are not antigens. They are referred to as haptens. When linked covalently to albumin, they act as antigens with the ability to generate antibodies to the hapten portion of the antigen complex (see Figure 12.1). To form the hapten-albumin complex, hydroxyl groups or ketone groups of the ligand are linked covalently to the NH_2 group of lysines in the albumin molecule. It is said that antibodies are "farsighted," i.e., they can recognize with exquisite specificity the portion of the hapten that is farthest removed from the protein carrier to which it is linked. For example, antibodies generated against the T_3-albumin complex cross-react to an extent less

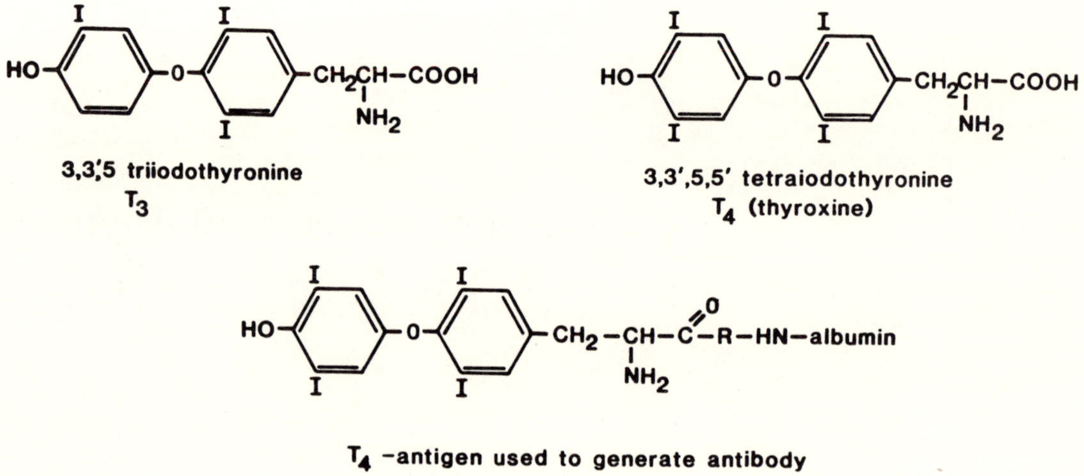

HO–C₆H₂I₂–O–C₆H₃I–CH₂CH(NH₂)–COOH

3,3',5 triiodothyronine
T_3

HO–C₆H₂I₂–O–C₆H₂I₂–CH₂CH(NH₂)–COOH

3,3',5,5' tetraiodothyronine
T_4 (thyroxine)

HO–C₆H₂I₂–O–C₆H₂I₂–CH₂–CH(NH₂)–C(=O)–R–HN–albumin

T_4 –antigen used to generate antibody

Figure 12.1. Structures of T_3 and T_4 (top). Structure of the T_4 antigen used to generate T_4 antibodies (bottom).

than 1/10,000 with the antibody-to-T_4-albumin complex, even though the difference between T_3 and T_4 is one I atom on the thyronine nucleus.

Radioimmunoassay of thyroxine

Thyroxine bound covalently to albumin is injected into an animal to challenge its immune system. Over time an antibody formed to the thyroxine-albumin complex can be isolated from the blood of the animal. The antiserum, or antibody, is then mixed with a radioactive derivative of thyroxine and with human serum or standard solutions of thyroxine. The thyroxine in the serum or the standard solutions and the radioactive thyroxine compete equally for the binding site on the antibody. That is, the antibody binds either the natural thyroxine or the labeled thyroxine with equal affinity. Therefore the greater the amount of naturally occurring thyroxine present in the serum or the standard solutions, the lower the amounts of radioactive thyroxine that will be bound to the antibody complex. One must have a precise procedure for the separation of the hormone-antibody complex from the free hormone and a method to determine the amount of radioactivity bound to the complex.

A standard curve is constructed (see Figure 12.2). The percentage of radioactive thyroxine bound to the antibody after incubation with the unknown serum is determined, and with the aid of the standard curve the amount of thyroxine in the serum can be estimated as shown.

Clinical Applications

Measurement of Bound and Free Thyroxine

The thyroid gland secretes two hormones, thyroxine (T_4) and triiodothyronine (T_3) (Figure 12.1), which have a major influence on regulating human metabolism. The symptoms of thyroid dysfunction are sometimes not readily discerned. Therefore the availability of laboratory tests is of major importance in the diagnosis of thyroid disease.

Most T_4 and T_3 circulate in the blood bound to carrier proteins, principally (70%) thyroid-binding globulin (TBG), and to a lesser extent thyroid-binding prealbumin (TBPA) and serum albumin. The binding of thyroid hormone to serum proteins is significant: 99.96% of circulating T_4 and 99.6% of circulating T_3. The small amount of unbound or free hormone concentration is biologically effective and correlates with the degree of thyroid hormone action. Detailed procedures are available to estimate the concentration of free hormone. However, most clinical laboratories measure total (unbound and bound) T_4 in the serum. This measurement is the routine method of choice for most clinical situations since it usually correlates very well with the amount of free hormone.

Conditions that Alter Serum Levels of Total Thyroxine

Some examples of conditions that result in alterations in the amounts of total thyroxine in the serum are:

hyperthyroidism, high levels of thyroid-binding globulin (e.g., in pregnancy), thyrotoxicosis factitia, hepatitis, obesity, hypothyroidism, low levels of thyroid binding globulin, chronic liver disease, loss of protein from the gastrointestinal tract, and panhypopituitarism.

In some conditions high or low amounts of total thyroxine in the serum are not indicative of hyperthyroidism or hypothyroidism. The total thyroxine in the serum is a function of both the amount of thyroxine-binding globulin and also the rate of secretion of thyroxine from the thyroid gland. The accurate diagnosis of hyper- or hypothyroidism requires a clinical test, which can be used to indicate the contribution of high or low amounts of TBG in the serum to the high or low amounts of total thyroxine. The physician needs to know if the high or low amounts of total thyroxine are caused by high or low rates of secretion of thyroxine by the thyroid gland.

A procedure used in the clinic to estimate the contribution of the amount of TBG to the total thyroxine levels is the T_3 resin uptake test. Serum from the patient is mixed with a small or trace amount of ^{125}I-labeled T_3. The labeled T_3 will bind tightly to the TBG in the serum. An equilibrium is established between the free and TBG-bound ^{125}I-labeled T_3. Therefore the amount of ^{125}I-labeled T_3 added to the serum that binds to the TBG will be directly proportional to the amount of TBG present in the blood. The amount of labeled T_3 that is not bound to the TBG is determined by adding a resin to the serum that will trap or absorb the T_3 *not* bound to the TBG. The amount of labeled T_3 bound to the resin will be *inversely* proportional to the level of TBG present. This test, in other words, allows us to determine whether the patient has high or low amounts of TBG in his or her serum. It also enables us to consider the contribution of such levels to the total T_4 levels; we can determine that the high or low levels of total T_4 in the serum are the result of high or low rates of its secretion from the thyroid gland. The combination of the total T_4 RIA test and the T_3 resin uptake test permits the accurate diagnosis of hyperthyroid and hypothyroid conditions.

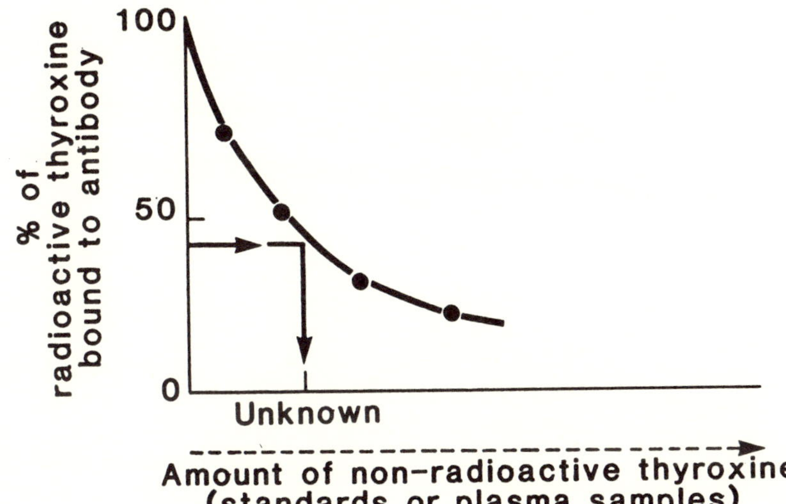

Figure 12.2. The use of a standard curve in the RIA of T_4.

Procedures

The thyroxine radioimmunoassay kits were purchased from Nuclear Diagnostics, Inc. A mixture of thyroxine specific antiserum and ^{125}I-labeled thyroxine is incubated with the serum samples and standards (unlabeled thyroxine) for one hour. During this incubation period the unlabeled thyroxine (from samples or calibration standards) competes with ^{125}I-labeled thyroxine for binding sites on the antibody complex. When the incubation period is complete, the antibody-thyroxine complex is packed by centrifugation, and the supernatant solution is immediately decanted and discarded. The radioactivity in the precipitate is determined with a gamma counter.

The following four reagents are used in the experiment.

1. Thyroxine (T_4) antiserum (Rabbit). Store at 2-8°C.

Components	*Concentrations*
PBS buffer	pH 7.2
T_4 antiserum (rabbit)	
Bovine gamma globulin	1.5%
Sodium azide	0.03% as a preservative

2. Thyroxine (T_4) ^{125}I reagent. Store at 2-8°C.

Components	*Concentrations*
sodium salicylate	1.56%
thyroxine (T_4) ^{125}I	$< 10\ \mu$Ci
tris-HCl buffer	0.12 M, pH 7.8
sodium azide	0.02% as a preservative

3. Standards in bovine or human serum, containing 0.1% sodium azide as a preservative. Store at 2–8°C. Refer to standard vial for thyroxine concentration.

a. Zero
b. Approximately 2.0 μg%
c. Approximately 4.0 μg%
d. Approximately 6.0 μg%
e. Approximately 12.0 μg%
f. Approximately 24.0 μg%

4. Polyethylene glycol (PEG), 25% solution. Store at 2-8°C.

Students should work in groups of four.

1. Label 11 assay tubes (12 × 75 mm plastic) for the reference control serum, the student serum, and the thyroxine standards.
2. Pipette 0.4 ml (400 μl) of ^{125}I reagent into each test tube.
3. Pipette 0.01 ml (10 μl) of serum samples, standards, and controls into the labeled test tubes and mix.
4. Pipette 200 μl of T_4 antiserum to thyroxine into each test tube. Shake vigorously for 10–15 sec.
5. Incubate the test tubes at room temperature (15–30°C) for one hour.
6. Add 1.0 ml of cold PEG reagent to all test tubes. NOTE: for optimum precipitation of the antigen/antibody complex, the PEG reagent must be at refrigerator temperature when used. *Do not permit reagent to warm to room temperature*. For best results, remove from refrigerator right before use.
7. Vortex all test tubes for 5–10 seconds. The mixture should appear evenly opaque or cloudy. Insufficient vortexing may cause incomplete precipitation, resulting in poor precision.
8. Centrifuge all test tubes in a refrigerated centrifuge for 10 minutes at 3,000 rpm (1,300–1,500 g's). Remove from centrifuge and immediately decant the supernate by inversion of the tube in a smooth gradual motion. Remove the last drop of supernate by touching the lip of the test tube to an absorbent surface to allow liquid to drain from tube. The adhesive nature of the precipitate prevents it from dislodging and draining out of the test tubes.
9. Calculate as follows:

$$\frac{\text{net counts of sample or standards}}{\text{net counts of the 0 standard}} \times 100 = \%\text{ bound (relative to 0 standard).}$$

Plot % bound vs concentration of the standards and calculate unknown samples. (See Figure 12.2).

Expected values: approximately 5–13 μg/dl

Selected References

Berger, S., and Quinn, J. L. 1976. Thyroid Function. In *Fundamentals of Clinical Chemistry*, ed. N. W. Tietz, 2nd ed., pp. 824–848. Philadelphia, Saunders.

Howanitz, P. J., and Howanitz, J. H. 1979. Evaluation of Endocrine Function. In *Todd-Sanford-Davidsohn: Clinical Diagnosis and Management by Laboratory Methods*, ed. J. B. Henry, 16th ed., Vol 1, pp. 402–476. Philadelphia, Saunders.

Markowitz, H., and Jiang, Nai-Siang. 1976. Immunochemical Principles. In *Fundamentals of Clinical Chemistry*, ed. N. W. Tietz, 2nd ed., pp. 274–297. Philadelphia, Saunders.

Steffes, M. W., and Oppenheimer, J. H. 1979. The Laboratory Evaluation of Thyroid Function. *Minnesota Medicine* 62:162–164.

Problems

Standards		*cpm* ^{125}*I-T*$_4$	
A 0	μg/dl		
B	μg/dl		
C	μg/dl		
D	μg/dl		
E	μg/dl		
F	μg/dl		
Sample		*cpm* ^{125}*I-T*$_4$	*μg/dl*
Control serum			
Student serum			
Student serum			
Student serum			
Student serum			

1. Record the counts per minute obtained from the above standard or serum samples. Calculate the amount of thyroxine in all serum samples.

2. Why should the data for the standard curve always be obtained simultaneously with the data for the unknown sample?

3. Give two reasons why T_4 levels in serum by this assay are not affected by serum T_3 levels.

4. The binding of T_4 by thyroxine-binding globulin is inhibited by the use of a TBG blocking agent (sodium salicylate) in the procedure described in the text. What would be the effect on the measured T_4 levels if this reagent was not incorporated into the procedure?

5. Radioimmunoassays are generally very simple to perform. However, radioimmunoassays have a larger technical error than many other analytical procedures. How can the technical error be kept to a minimum?

6. Explain why values for a hormone determined by radioimmunoassays might be at variance with the values obtained by biological assays.

7. The active portion of thyroid hormone in serum is the unbound fraction. Why can inferences be made about the clinical status by measuring the total concentration of thyroid hormone?

8. Explain why measurement of the total serum thyroxine may *not* correlate with the clinical status of the patient and give two specific examples.

13 Clinical Analysis of Serum Electrolytes

Charles W. Carr, Robert J. Roon, and John F. Van Pilsum

One of the major medical advances in this century has been the understanding and the management of fluid and electrolyte balance. Many pathological conditions in the human result in changes in the electrolyte composition of the extracellular fluid. Alterations in these components are often lethal and thus rapid procedures are required for the analysis of body fluids and remedial treatment of the imbalance. Procedures are presently available for the rapid determination of Na^+, K^+, Ca^{2+}, Cl^-, pCO_2, pO_2, pH, and bicarbonate.

Principles

Analysis of Serum Electrolytes

The normal values of serum electrolytes are:

	meq/liter
Na^+	136–145
K^+	3.5–5.0
Ca^{2+}	4.3–5.3
Mg^{2+}	1.5–2.5
Cl^-	100–106
HCO_3^-	23–27
HPO_4^{2-}/ $H_2PO_4^-$	1.7–2.6

The concentration of the following electrolytes are routinely determined in the clinical laboratory for the purpose of diagnosis and management of many diseases.

Sodium and Potassium

Sodium and potassium levels are often determined by flame photometry. With the flame photometer the concentrations of Na^+ and K^+ are measured by determining the intensity of the color produced when a solution of Na^+ and K^+ salts is aspirated into a flame. Na^+ and K^+ each have a characteristic emission spectrum when heated to a high temperature, the wave length of the emission being in the visible range. For sodium it is 589.0 nm (yellow), and for potassium it is 766.5 nm (purple). With appropriate filters the intensities of both sodium and potassium emissions can be measured in a single determination.

The concentration of sodium can also be determined by the use of a sodium ion-selective glass electrode. This type of electrode permits the rapid, accurate measurement of the activity of a given ion in a solution. The instrument consists of two electrodes that are immersed in the solution being tested and a potentiometer that measures the difference in potential between the electrodes. One of the electrodes is a reference electrode that has a constant potential. The other is a glass membrane electrode, the potential of which varies with the activity of the sodium ion solution.

Potassium levels can be determined in this manner with the use of a potassium-selective membrane electrode. These electrode methods are employed in some recently developed multichannel analyzers.

Calcium

The calcium concentration is usually measured by atomic absorption. The principle of atomic absorption spectroscopy is similar to that of the flame photometer. A special cathode tube produces the characteristic emission spectrum of calcium (422.7 nm), and this light is passed through the vaporized calcium-containing sample. Calcium atoms in the ground state absorb the incoming radiation, the degree of absorption being proportional to the concentration of calcium in the vapor. With an appropriate detection system the amount of absorption can be measured and compared with that of solutions of known calcium concentrations.

Calcium levels can also be determined by titration with the chelating agent EDTA at an alkaline pH. A calcium indicator is used to indicate the point at which all the calcium is complexed with EDTA. Other methods include precipitation with chloranilic acid and flame photometry.

Bicarbonate

Levels of HCO_3^- can be measured by direct titration. To the plasma sample containing bicarbonate, an excess of HCl is added, i.e., an amount of HCl in excess of that needed to react with all the bicarbonate and convert it to CO_2. The CO_2 is removed, and the remaining HCl is titrated with NaOH until the pH of the plasma sample is brought back to its original value.

In some laboratories total CO_2 is determined since it serves as an approximation of HCO_3^-. (HCO_3^- = total CO_2 − 1.2 mmol/l. The 1.2 mmol/l = correction factor for the estimated level of H_2CO_3.) All forms of CO_2 (CO_2, H_2CO_3, and HCO_3^-) are converted to CO_2 in the gaseous state. The CO_2 (g) is measured by a pCO_2 electrode, a microgasometer, or some other method.

Chloride

The chloridometer is an apparatus that involves an electrometric analysis of chloride in blood and urine. A microammeter measures the current between two silver electrodes while silver ion is being added to the solution. The end point of the reaction (when all the chloride is converted to AgCl) is detected by a sudden increase in the current.

Analyses for Blood Gases and pH

The normal values of blood gases and pH are:

	Arterial blood (mm Hg)	*Venous blood (mm Hg)*
pCO_2	35–45	38–50
pO_2	95	5–40
pH	7.35–7.45	7.33–7.43

The *pH* of blood or plasma is determined by the use of a hydrogen ion selective glass electrode.

pO_2 is measured by determining the amount of current produced by the reduction of O_2 present in the sample. The pO_2 electrode used is a complete electrochemical cell. The cathode is covered with a gas permeable membrane to make the electrode specific for O_2 reduction.

pCO_2 is usually measured by a pCO_2 electrode (a specialized pH electrode). The CO_2 passes through a gas-permeable membrane, causing a change in the pH of a solution. The change in pH causes a change in potential.

pCO_2 can also be estimated if the pH and bicarbonate concentration of a solution are known. An equilibrium equation relating pH, CO_2 tension, and concentration of bicarbonate can be derived from the Henderson-Hasselbach Equation as follows:

$$pH = pK_a + \log \frac{[HCO_3^-]}{[H_2CO_3]}.$$

The pK_a = 6.1 and, under physiological conditions,

$$[H_2CO_3] = \alpha pCO_2 \text{ so:}$$

$$pH = 6.1 + \log \frac{[HCO_3^-]}{\alpha pCO_2}.$$

When (HCO_3^-) is expressed in terms of millimoles/liter and pCO_2 is given in terms of mm of Hg, $\alpha = 0.03$. The above equation can then be converted to:

$$pH = 7.62 + \log (HCO_3^-) - \log pCO_2$$

Chloride Shift

The principal mechanism for the transport of CO_2 in the blood is in the form of bicarbonate ions. The formation of bicarbonate by the reaction of CO_2 with the plasma buffers accounts for only a few percent of the total bicarbonate transport. Nearly all of the bicarbonate formed in blood passing through the tissues is the result of a complex interaction of CO_2 with hemoglobin (see Figure13.1).

The following events occur as bicarbonate and chloride ions interact with red blood cells.

In the capillaries of tissues oxygen is released from oxyhemoglobin (HbO_2^-) and the proton dissociation for certain amino groups in globin is diminished (the affinity for protons is increased).

Also in the capillaries carbon dioxide that has entered the red blood cell becomes converted to carbonic acid (H_2CO_3). The carbonic acid dissociates to bicarbonate (HCO_3^-) and donates a proton to the deoxyhemoglobin (isohydric shift). Bicarbonate moves from the red cell by passive diffusion and is replaced with an equivalent amount of chloride to maintain electro-neutrality (chloride shift).

The red cells are transported to the capillaries of the lungs, where oxygen enters, binds to the globin, and

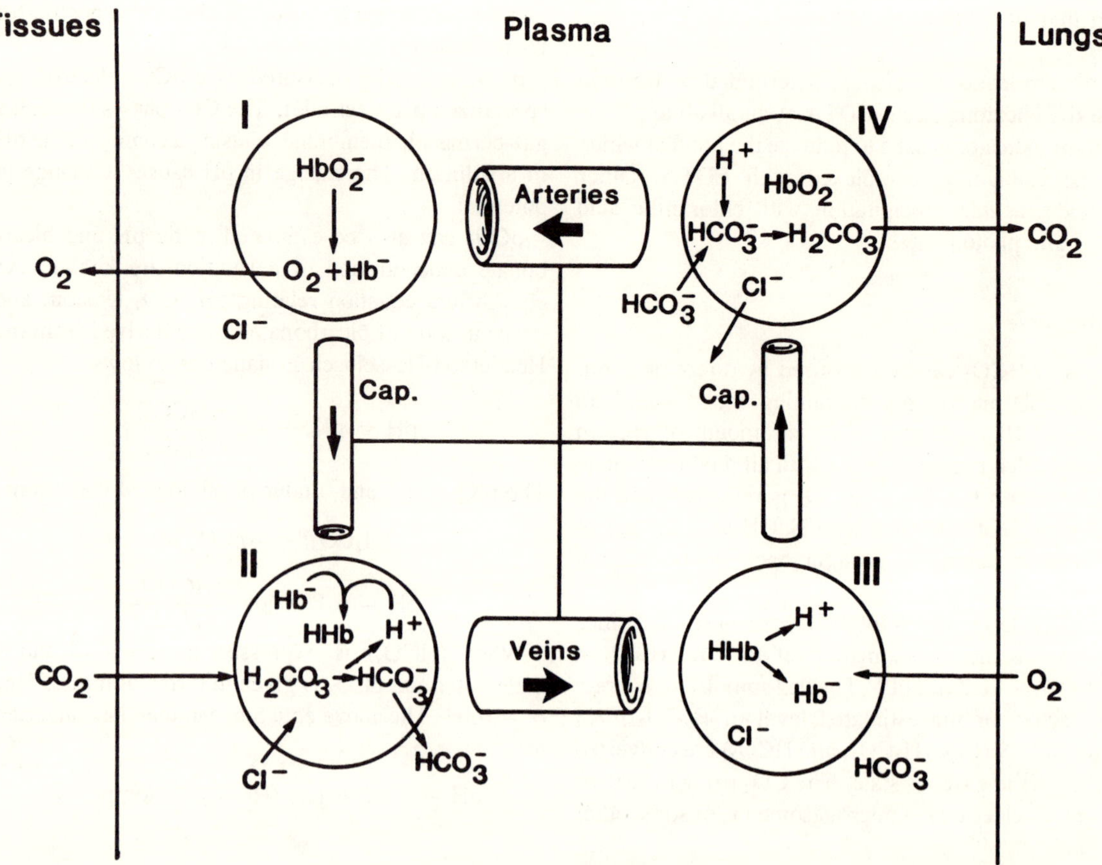

Figure 13.1. The chloride shift between plasma and red blood cells

increases the proton dissociation (weakens the affinity for protons).

The protons released from the globin react with bicarbonate and shift the equilibrium toward the formation of carbonic acid and carbon dioxide, thus allowing additional bicarbonate to diffuse into the red cell. Chloride ions exit from the cell to maintain electro-neutrality.

The net result of these processes is that the concentration of Cl^- in venous blood plasma is lower than that found in arterial blood plasma.

Clinical Applications

Alterations in Serum Electrolytes that Occur in Various Disease States

Sodium

The volume of the extracellular fluid is directly related to the total available sodium. This relationship is not fully understood, but it is clear that alterations in the plasma sodium concentration are associated with many of the abnormal circumstances in which a change in extracellular volume occurs. Low sodium levels in serum are found in diabetes insipidus, metabolic acidosis, Addison's disease (a decreased secretion of corticosteroids), diarrhea, and so on. High sodium levels in serum are indicative of Cushing's syndrome (an increased production of mineralocorticosteroids), severe dehydration, diabetic coma, and so on.

Chloride

Chloride is the principal anion of the extracellular fluid, and changes in its concentration usually accompany alterations in sodium concentration. Its analysis is especially useful in cases of acid-base disturbances. Low chloride values in serum are associated with Addisons's disease, metabolic acidosis, and prolonged vomiting; high values, with dehydration and congestive heart failure.

Potassium

The potassium level of the extracellular fluid is not as well regulated as the sodium level; nevertheless the maintenance of the normal potassium level in plasma is important. High potassium concentrations affect the heart first. At about 6mM potassium, changes occur in the electrocardiogram, and at 9—10 mM potassium the heart may stop. Increases in plasma potassium occur in hypoadrenocorticalism, severe dehydration, anuria, renal failure due to shock, and urinary obstruction. A decrease in plasma potassium leads to extreme muscular weakness, myocardial degeneration, and paralysis. This decrease may be caused by diuresis, diarrhea, acid-base imbalance, or insufficient intake.

Calcium

The calcium level of the extracellular fluid is very well regulated. Such substances as parathyroid hormone, calcitonin, and Vitamin D all play a direct role in this regulation. A decrease in plasma calcium leads to tetany, and an increase produces a comatose condition that may lead to respiratory or cardiac failure. Hypoparathyrodism is one of the principal causes of low plasma calcium, and hyperparathyroidism causes increased plasma calcium.

Bicarbonate

Changes in the bicarbonate level of the plasma are usually associated with acid-base disturbances, especially those of metabolic origin. Thus metabolic acidosis produces a lowered concentration of plasma bicarbonate, and metabolic alkalosis is accompanied by an increase in the plasma bicarbonate.

Magnesium

Increases in serum magnesium levels are found in dehydration, severe diabetic acidosis, and Addison's disease. Decreased magnesium levels in serum are indicative of the malabsorption syndrome, acute pancreatitis, chronic alcoholism, chronic glomerulonephritis, and other diseases.

Phosphate

Increased phosphorus levels in serum are found in hypervitaminosis D, hypoparathyroidism, and renal failure; low levels, in rickets, hyperparathyroidism, and the Fanconi syndrome—a disease associated with a defect in the reabsorption of phosphorus from the glomerular filtrate.

Procedures

Procedures for the Analysis of Plasma Electrolytes

In these experiments students become familiar with the concentration units, normal ranges, and some of the pathological conditions that result in abnormal electrolyte values. The electrolytes to be determined are Na^+, K^+, and Cl^-.

Sodium and Potassium

The use of a Coleman Jr. flame photometer is demonstrated. Students then determine the sodium and potassium concentrations of their unknown samples of plasma or serum.

Sodium analysis: Pipette 0.50 ml of sample into a 100 ml volumetric flask and fill to within 5 ml of the mark with distilled water. Add 2.0 ml of Acationox solution to the flask and dilute to the mark with water. Mix well and bring to flame photometer for analysis.

Potassium analysis: Pipette 0.50 ml of sample into a 25 ml volumetric flask and fill to within 10 ml of the mark with water. Add 5.0 ml of Acationox sodium chloride reagent to the flask and dilute to the mark. Mix well and bring to the flame photometer for analysis.

Chloride

The use of the chloridometer is demonstrated with standards and blank solutions. The unknown serums are analyzed for their content of chloride.

Chloride analysis: With an Eppendorf pipette, add 100μl of sample to the special titration vials for chloride. Four ml of nitric acid-acetic acid-PVC reagent is added to the vial, and the amount of Cl^- in the solution is determined with the aid of the chloridometer.

Record values for potassium, sodium, and chloride (see Sample Worksheet 13.1).

Estimation of CO_2 Tension of Alveolar Air

This experiment illustrates in a relatively simple manner *the relationship between the pH of a solution, the bicarbonate concentration and pCO_2 of the gas in equilibrium with the solution.* Our knowledge of acid-

Sample Worksheet 13.1. Determination of Serum Electrolytes

Electrolyte	Standard		Unknown	
	Reading (mV)	meq/1	Reading (mV)	meq/1
K^+				
Na^+				
Cl^-				

base balance is dependent on an understanding of this basic concept. To carry out the experiment the student equilibrates a solution of known bicarbonate concentration with his or her own alveolar air. This procedure also shows that alveolar air has a higher pCO_2 than expired air.

Procedure: Three bicarbonate solutions are available; 20, 30, and 40 m*M*. Place about 25 ml of one of the bicarbonate solutions in a 100 ml beaker. After rinsing the electrodes carefully, immerse the two electrodes of a pH meter into the beaker. The electrodes will previously have been fitted with a rubber stopper so that only a small opening remains for the escape of air.

Measure the pH of the bicarbonate solution. Place a piece of rubber tubing over the glass tubing, which is also fitted into the rubber stopper. At the end of a normal expiration the rubber tubing is placed in the mouth, and the residual air from the lungs is forced out and bubbled through the solution. Repeat this forced expiration of air through the solution several times, while observing the pH of the solution. This procedure is continued until the pH of the solution no longer changes. Repeat the experiment with the other bicarbonate solutions, and record the equilibrium pH values (see Sample Worksheet 13.2). *Calculate the pCO_2 for each of the three solutions, using the formula given under Principles.* Assume that the bicarbonate level does not change from its original concentration.

Sample Worksheet 13.2. Calculation of CO_2 Tension of Alveolar Air

HCO_3^-	pH	pCO_2
20 mM		
30 mM		
40 mM		

Measurement of Chloride Shift

The reality of the chloride shift is illustrated in this experiment. A sample of whole blood is divided into two portions, one sample being equilibrated with O_2 and the other with CO_2. Plasma chloride analyses are then carried out on the two samples.

Procedure: Place 5 ml samples of whole blood into each of two *250 ml* Erlenmeyer Flasks. Blow CO_2 into one flask, swirling the flask so that a large surface area of blood is exposed to the gas. Do *not* bubble the gas through the blood. After four to five minutes of this treatment, pour the blood into a 12 ml centrifuge tube. Overlay the blood with a layer of mineral oil about 1/4″ deep. Treat the second flask with O_2 in the same manner and transfer the blood to a second centrifuge tube.

Centrifuge both tubes at full speed for 10 minutes. Draw off the plasma from each tube, using a Pasteur pipette so that the plasma layer can be carefully separated from the oil and cell layers. Carry out duplicate chloride analyses as directed on p. 95 and record (see Sample Worksheet 13.3).

Sample Worksheet 13.3. Effects of Carbon Dioxide on Plasma Chloride

Gas	Cl^- (meq/1)
O_2	
CO_2	

Selected References

Siggaard-Andersen, O. 1976. Electrochemistry. In *Fundamentals of Clinical Chemistry*, ed. N. W. Tietz, 2nd ed., pp. 135–153. Philadelphia, Saunders.

Siggaard-Andersen, O. 1976. Blood Gases. In *Fundamentals of Clinical Chemistry*, ed. by N. W. Tietz, 2nd ed., pp. 854–873. Philadelphia, Saunders.

Smith, E. L., Hill, R. L., Lehman, I. R., Lefkowitz, R. J., Handler, P., and White, A. 1983. *Principles of Biochemistry: Mammalian Biochemistry*, 7th ed., pp. 100–207. New York, McGraw-Hill.

Tietz, N. W. 1976. Electrolytes. In *Fundamentals of Clinical Chemistry*, ed. N. W. Tietz, 2nd ed., pp. 873–944. Philadelphia, Saunders.

Tietz, N. W., and Siggaard-Andersen, O. 1976. Acid-Base and Electrolyte Balance. In *Fundamentals of Clinical Chemistry*, ed. N. W. Tietz, 2nd ed., pp. 945–974. Philadelphia, Saunders.

Problems

1. Explain the low serum concentrations of sodium found in diarrhea and Addison's disease.

2. Explain the low serum concentrations of chloride found in Addison's disease, metabolic acidosis, and prolonged vomiting.

3. Explain the low serum levels of potassium resulting from diarrhea, metabolic acidosis, and a low potassium diet.

4. In serum the difference between the sum of (Na^+) + (K^+) and (Cl^-) + (HCO_3^-) is sometimes referred to as the anion gap. In normal serum this difference is about 12–15 meq/l. What is the explanation for the anion gap?

5. For a sample of plasma the sodium is 345 mg% and the potassium is 9.5 mg%. What are these concentrations in meq/1 and in micromoles per 100 ml? (Atomic weights are 23 for sodium and 39 for potassium.)

6. Given a pCO_2 of 40 mm tension in alveolar air, what is the mM concentration of (H_2CO_3), in plasma?

7. If arterial blood plasma has a pH of 7.40 when it is equilibrated with a pCO_2 of 32 mm, what is its bicarbonate concentration?

8. The pK_a for $H_2PO_4^{-1}$ in plasma is 6.8. If the plasma pH is 7.4, what is the ratio of $HPO_4^{-2}/H_2PO_4^-$?

9. Explain why acidosis is produced in diabetes and starvation.

14 Lecithin/Sphingomyelin Ratio of Amniotic Fluid

Maureen A. Scaglia and John F. Van Pilsum

A leading cause of death in premature babies is respiratory distress syndrome (RDS). RDS is caused by lung immaturity due to a lack of lung surfactant, a major component of which is a phospholipid called lecithin. Fluid containing lung lecithin flows from the lungs of the fetus into the amniotic fluid. By determining the amount of lecithin relative to the amount of sphingomyelin in amniotic fluid, an estimate of lung maturity can be made. Determination of the lecithin/sphingomyelin ratio involves extraction and precipitation of the phospholids, their separation by thin-layer chromatography, and comparison of sample spots with those of the standards. The results of this method aid the physician in determining the optimal time for delivery.

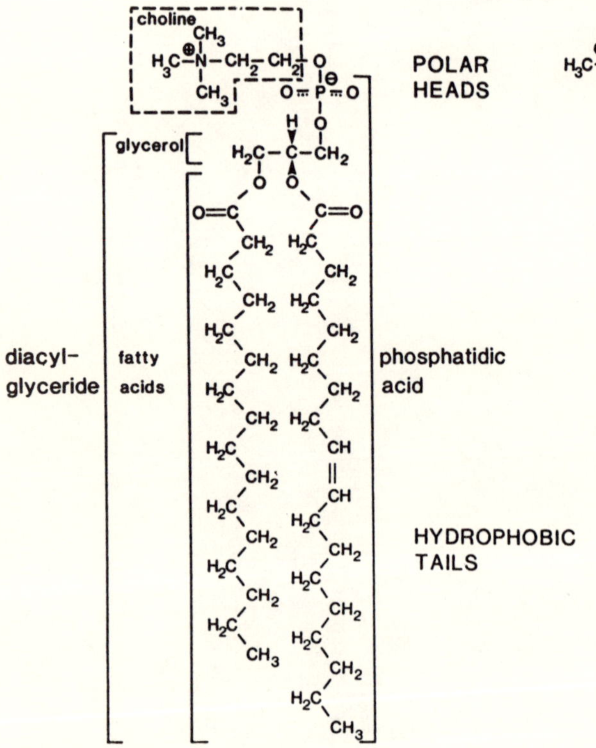

choline
glycerol
diacyl-glyceride
fatty acids
phosphatidic acid
POLAR HEADS
HYDROPHOBIC TAILS

Figure 14.1. Structure of a lecithin (a phosphatidylcholine) indicating the components utilized in its synthesis. (The fatty acids shown are palmitic and oleic.)

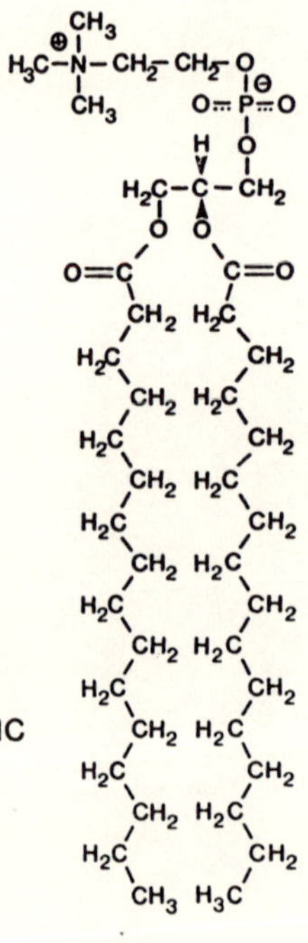

Figure 14.2. Structure of dipalmitoyl lecithin (dipalmitoyl phosphatidylcholine.)

Principles

Normal lung function depends on the presence of surfactant, which lines the inside of the alveoli. The walls of the alveoli are not strong enough to maintain their shape, after expiration, against the surface tension of water. Surfactant coats the walls of the alveoli and prevents their collapse. The major components of surfactant are a group of phospholipids. These compounds have a hydrophobic part (long-chain fatty acids in ester linkage) and a hydrophylic part (the phosphatidyl component). This amphipatic structure promotes concentration of the molecules at air-liquid interfaces, with the hydrophobic end in the air and the hydrophilic part in the aqueous medium. The surfactant thus reduces surface tension, allowing the alveoli to remain partially inflated after expiration. Without surfactant, the alveoli collapse, breathing becomes difficult, and RDS occurs.

The lecithins (phosphatidylcholines) are produced in the lung and other parts of the body. They are phospholipids composed of a 1,2 diacylglyceride (α, β diglyceride) and a phosphodiester bridge linking the glycerol with a choline base (see Figure 14.1). Dipalmitoyl phosphatidylcholine (dipalmitoyl lecithin) comprises 80% of phospholipid surfactant. The diacylglyceride portion of this phosphatidylcholine contains two palmitic acid residues (a fatty acid of 16 carbon atoms and no double bonds). Since they are fully saturated, the long fatty acid chains are not bent (see Figure 14.2) and can be packed tightly, allowing for high surface activity. Dipalmitoyl phosphatidylcholine is synthesized in type II alveolar epithelial cells and stored in their lamellar bodies. These cells appear at about 24 weeks gestation and increase in number until 32 weeks. At that time dipalmitoyl phosphatidylcholine appears in the lung and amniotic fluid.

Synthesis of dipalmitoyl (disaturated) phosphatidylcholine is thought to proceed as described in Figure 14.3. In the lung, most of the dipalmitoyl phosphatidylcholine (dipalmitoyl PC) is not synthesized directly from cytidine diphosphate (CDP) choline and 1,2 diacylglyceride. Instead, as gestational age increases, preformed unsaturated phosphatidylcholine (lecithin) is deacylated by phospholipase A_2 to lysophosphatidylcholine (PC minus one fatty acid). Lysophosphatidylcholine is then transformed by reacylation or transacylation to dipalmitoyl phosphatidylcholine.

Tests for presence of lung surfactant fall into two major groups: those that measure the physical properties of surfactant and those that measure the chemical components. Examples of the first group include direct measurement of amniotic fluid surface tension; fluorescence polarization (based on the relationship between fluorescence polarization, microviscosity, and surface tension); and the Foam (''Shake'') Stability Test.

Of the second group, the lecithin/sphingomyelin ratio is the most commonly used test for determining fetal lung maturity. After extraction and precipitation of the phospholipids; unknown samples and standards are applied to thin-layer chromatography plates. Following separation, visualization of the spots is achieved by staining or charring. The concentration of lecithin and sphingomyelin is compared and quantitated. Methods of quantitation include densitometry, phosphorus determination, and area measurement. An L/S ratio of less than 2:1 is likely to be associated with RDS.

CDP-choline + 1-palmitoyl-2-unsaturated diacylglycerol $\xrightarrow[2\ Mg^{++}]{\text{cholinephosphotransferase}}$ 1-saturated (palmitoyl)-2-unsaturated acylPC + CMP

1-palmitoyl-2-unsaturated acylPC $\xrightarrow{\text{phospholipase } A_2}$ 1-palmitoyl-2-lysoPC + an unsaturated fatty acid

2 1-palmitoyl-2-lysoPC $\xrightarrow{\text{lysoPC:lysoPC acyltransferase}}$ dipalmitoylPC + Glycero-3-phosphorylcholine

OR

1-palmitoyl-2-lysoPC + palmitoyl-CoA $\xrightarrow{\text{acyl:CoA lysoPC acyltransferase}}$ dipalmitoylPC

Figure 14.3. Synthesis of dipalmitoyl phosphatidylcholine (PC = phosphatidylcholine).

Sphingomyelins are phosphoplipids that resemble phosphatidylcholines in shape. The chemical structure can be seen in Figure 14.4. Instead of glycerol and one of the fatty acids, sphingomyelin molecules contain sphingosine, a long-chain dihydroxy amine. Sphingomyelins serve as structural lipids in nervous tissue. The concentration of sphingomyelin in amniotic fluid remains relatively constant during pregnancy, in contrast to lecithin levels which rise rapidly after 32–34 weeks gestation.

Other phospholipids beside lecithin and sphingomyelin are found in lung surfactant. Another major phospholipid is phosphatidyl glycerol (PG). Its concentration increases after 36 weeks gestation to 10% of total phospholipid concentration. Lung surfactant also contains phosphatidylethanolamine, phosphatidylinositol, and phosphatidylserine.

Other tests measure total phospholipid concentration, lecithin concentration, or palmitic acid concentration. Studies have shown the L/S ratio may be unreliable in certain cases, such as when the amniotic fluid is contaminated with hemoglobin, or in complicated pregnancies. In such cases, a newer technique of quantifying the amount of saturated phosphatidylcholine (SPC) offers an additional predictive value.

Clinical Applications

The lecithin/sphingomyelin ratio can aid the physician in determining gestational age and/or optimal time for delivery of high-risk pregnancies. In normal pregnan-

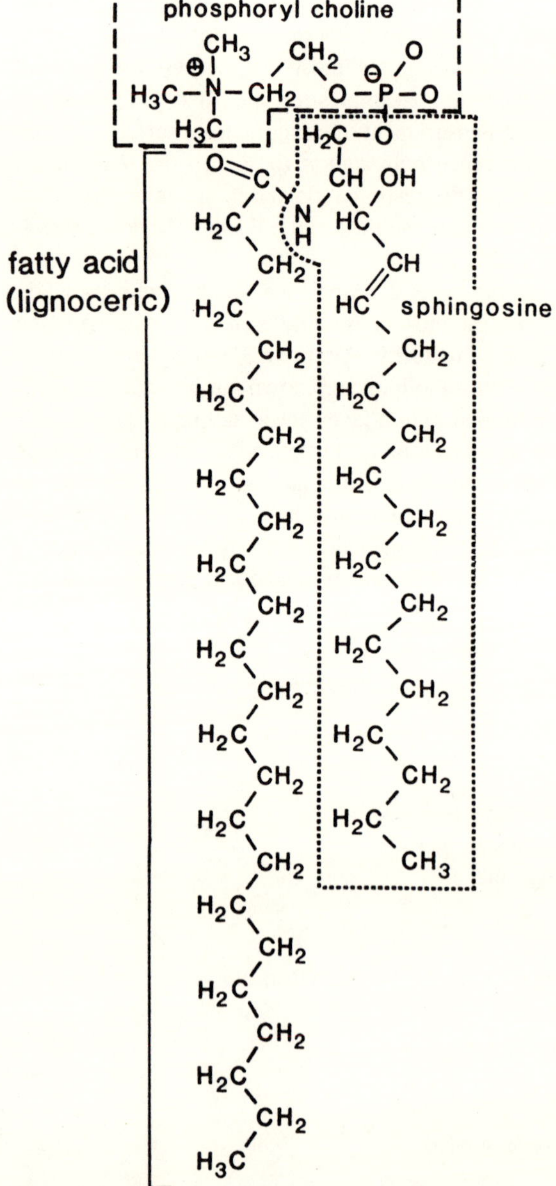

Figure 14.4. Structure of a sphingomyelin indicating the components utilized in its synthesis.

cies the L/S ratio is 1 or less until 30–32 weeks gestation, reaches 2 at about 35 weeks, and continues to increase until delivery. Studies have shown there is a 1.5% chance of an infant developing RDS with L/S ratios above 2.0 and a 78% chance below 1.5 (see Harvey and Parkinson, 1981). A ratio of 2 indicates lung maturity.

Some maternal diseases and conditions accelerate production of lung surfactant and an earlier L/S ratio of 2. Examples of these include severe toxemia, severe placental problems, some types of diabetes mellitus, and narcotic addiction. If delivered prematurely, these infants often do not develop RDS. Sharp increases in the L/S ratio may indicate acute stress because stored surfactant can be released during labor or under stressful conditions.

Some diseases, especially other types of diabetes mellitus, appear to slow the production of lecithin. In some diabetic pregnancies, there is an increased incidence of RDS associated with a L/S ratio of 2 or greater. The presence of phophatidyl glycerol or measurement of SPC, combined with an L/S ratio above 2, offers greater assurance of a safe delivery in a diabetic pregnancy.

Procedure

Lipids are extracted from amniotic fluid with methanol and chloroform. Phospholipids are then precipitated with acetone. These are separated by thinlayer chromatography. Visualization is achieved by staining in iodine vapor. The lecithin/sphingomyelin ratio can then be estimated.

1. Pipet 2 ml of each amniotic fluid sample into a 12-ml centrifuge tube. (The amniotic fluid has been previously centrifuged to remove cellular elements.)
2. Add 2 ml methanol to each sample and mix, using a Vortex mixer. Then add 4 ml chloroform. Cork tightly and mix by rapid inversion (15–20 ×).
3. Centrifuge for five minutes in a bench-type centrifuge at full speed. Remove upper solvent layer and protein precipitate by aspiration.
4. Add approximately 1 g Na_2SO_4 to each tube and vortex. Centrifuge for five minutes at full speed.
5. Using Pasteur pipettes, transfer each supernatant to a 15 ml centrifuge tube, avoiding floating water droplets.
6. Place tubes in a warm water bath (60°C) and evaporate contents to dryness under a stream of air.
7. Place the tubes on ice for five minutes, then add to each tube 0.75 ml of cold dry acetone (made by adding 5 g Na_2SO_4 to 1 qt acetone, mixing, and refrigerating). A white precipitate may form as the first drops are added. Return tubes to the ice for five minutes.
8. Centrifuge for 10 minutes at 3,000 rpm and 4°C. Decant and discard the supernatant. Place the tubes in the warm water bath for a few minutes to evaporate any remaining acetone.
9. Add 50 μl chloroform to the bottom of each tube. Using a capillary tube, transfer the sample to a spot 2 cm from the bottom of a TLC plate. (Use a 10 × 20 cm silica gel-coated glass TLC plate that has been drying for one hour at 80°C. Remove from oven right before use.)
10. Apply 10 μl of each standard solution (mixtures of lecithin and sphingomyelin in varying proportions)

to the plate using a microcapilliary pipette (see Figure 14.5).

11. Place the plate in a developing tank equilibrated with developing solvent (chloroform:methanol:H_2O, 65:25:4). Allow solvent to rise 10 cm. Remove the plate from the tank and let it dry.

12. Place the plate in a staining jar containing iodine crystals and develop until the spots are clearly visible.

13. Remove the plate and immediately compare the samples' spots with that of the standards. Estimate the lecithin/sphingomyelin ratio in each sample. Phosphatidyl glycerol may be present near the solvent front (see Figure 14.5). The spots will fade after 10 minutes. If necessary, return plate to staining jar for revisualization of spots.

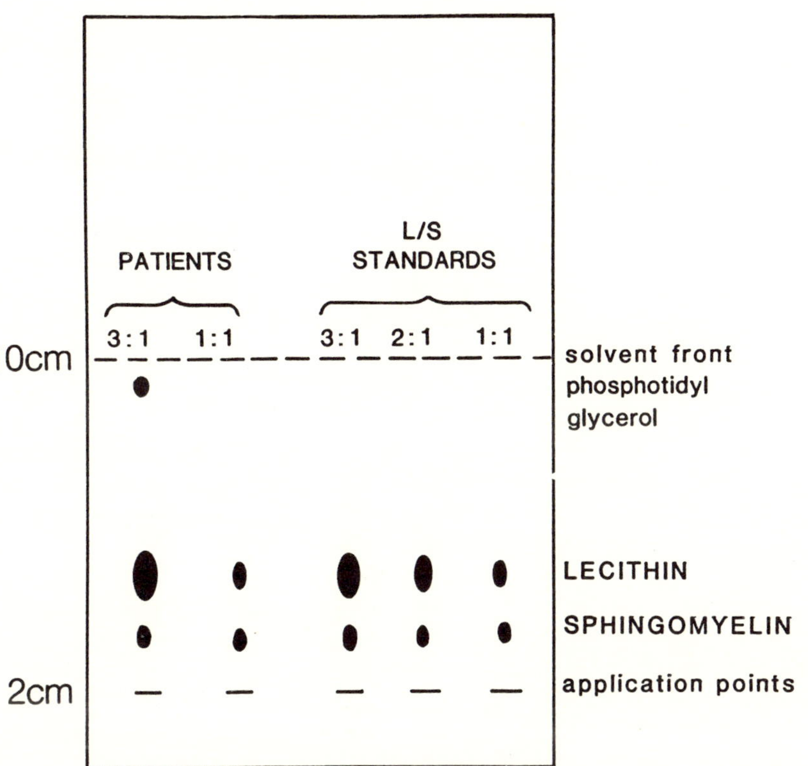

Figure 14.5. An example of chromatogram obtained from this experiment.

Selected References

Behrman, R. E. 1977. *Neonatal-Perinatal Medicine: Diseases of the Fetus and Infant*. pp. 14–21. St. Louis, Mosby.

Glew, R. H. 1982. Lipid Metabolism II: Pathways of Metabolism of Special Lipids. In *Textbook of Biochemistry with Clinical Correlations*, ed. T. M. Devlin, pp. 488–541. New York, Wiley and Sons.

Gluck, L., and Kulovich, M. V. 1973. Lecithin/Sphingomyelin Ratios in Amniotic Fluid in Normal and Abnormal Pregnancy. *Am. J. Obstet. Gynecol*. 115:539–546.

Harvey, D., and Parkinson, C. E. 1981. Prediction of the Respiratory Distress Syndrome. In *Laboratory Investigation of Fetal Disease*, ed. A. J. Barson, pp. 267–298. Bristol, Wright and Sons.

McGilvery, R. W. 1983. *Biochemistry: A Functional Approach*. pp. 203–215. Philadelphia, Saunders.

Torday, J., Carson, L., and Lawson, E. E. 1979. Saturated Phosphatidylcholine in Amniotic Fluid and Prediction of the Respiratory-Distress Syndrome. *N. Engl. J. Med*. 301:1013–1018.

Problems

1. Draw the chromatogram obtained from this experiment. What is the L/S ratio of the samples of amniotic fluid?

2. Why are phospholipids so important in the structure of membranes and as components of surfactant?

3. Why is dipalmitoyl phosphatidylcholine (lecithin) the most effective lung surfactant?

Appendix

Reagents

1. Hemoglobulin electrophoresis

   NaCl, 0.9%

   Tris-EDTA-borate buffer, 0.12 M, pH 9.2
   Dissolve 12.1 g tris (hydroxymethyl) aminomethane, 1.56 g disodium ethylenediaminetetraacetate,and 0.92 g boric acid in deionized H_2O and dilute to 1 liter. Adjust pH to 9.2 with 50% NaOH.

2. Plasma protein electrophoresis

   Barbital buffer, pH 8.6, ionic strength 0.075
   Dissolve 2.76 g diethylbarbituric acid and 15.45 g sodium barbital in deionized H_2O and dilute to 1 liter.

   Ponceau S fixative-dye
   Dissolve 0.5 g Ponseau S, 7.5 g trichloroacetic acid, and 7.5 g sulfosalicylic acid in deionized H_2O and dilute to 250 ml.

   Acetic acid rinse, 5%
   Dilute 50 ml glacial acetic acid to 1 liter with deionized H_2O.

   Bromphenol blue, 0.1%

3. LDH kinetics and isoenzymes assays

   Tris-HCl buffer, 83.4 mM, pH 7.4
   Dissolve 10.1 g tris (hydroxymethyl) aminomethane in about 900 ml of deionized H_2O. Adjust pH to 7.40 with 6 N HCl. Dilute to 1 liter with deionized H_2O.

   NADH, 4.8 mM
   Dissolve 34.5 mg NADH in 10 ml Tris-HCl buffer. Prepare fresh daily.

   Pyruvate, stock solution, 100 mM (11 mg/ml)
   Dissolve 2.200 g pyruvic acid, sodium salt, in tris-HCl buffer and dilute to 200 ml with buffer. Prepare fresh weekly. Store in refrigerator. Dilute stock for concentrations of pyruvate ranging from 0.2 mM to 20 mM.

   Lactic dehydrogenase
   Dilute 1 ml LDH (Sigma Chemical Co, type III, from bovine heart) to 100 ml with a 4% BSA-PO_4 buffer, pH 7.4. Prepare fresh daily.

   Phosphate buffer, 50 mM, pH 7.4
   Dilute 9.6 ml of 1.0 M KH_2PO_4 and 80.8 ml of 0.5 M Na_2HPO_4 to 1 liter with deionized H_2O. Check pH and adjust if necessary.

   Urea in phosphate buffer, 2.79 M, pH 7.4
   Dissolve 167.57 g urea in 50 mM phosphate buffer and dilute to 1 liter with buffer. Check pH and adjust if necessary.

   Hybridization solution, 1.0 M sodium phosphate, 10 mM β-mercaptoethanol, pH 7.0

   NaCl, 0.9%

4. LDH isoenzyme electrophoresis

   Barbital buffer, pH 8.6, ionic strength 0.075
   Dissolve 2.76 g diethylbarbituric acid and 15.45 g sodium barbital in deionized H_2O and dilute to 1 liter.

   Tris buffer, pH 8.4
   Dissolve 24.4 g tris (hydroxymethyl) aminomethane in about 900 ml of deionized H_2O. Adjust pH to 8.4 with 6 N HCl. Dilute to 1 liter with deionized H_2O.

   Tris-barbital buffer, pH 8.4
   Mix equal parts of tris and barbital buffers. Adjust pH to 8.4

   Lactic dehydrogenase stain
   Dissolve 0.210 g nitroblue tetrazolium, 0.020 g phenazine methosulfate, 1.00 g NAD, and 7.0 ml 60% lactic acid in 50 mM phosphate buffer and dilute to 100 ml with buffer. Store in brown glass bottle in the refrigerator for no longer than eight hours.

   Acetic acid rinse, 5%
   Dilute 50 ml glacial acetic acid to 1 liter with deionized H_2O.

5. Glucose by o-Toluidine and hexokinase methods

   o-Toluidine, 6% (v/v)
   Add 1.5 g reagent-grade thiourea to 940 ml of reagent-grade glacial acetic acid. Dissolve completely and then add 60 ml of o-toluidine. Mix well. Store in amber glass bottle. Age overnight before using.

   Glucose stock standard, 10.0 mg/ml
   Dissolve 1.000 g pure anhydrous glucose in and dilute to 100 ml with saturated benzoic acid (0.25%). Store in refrigerator. Dilute stock standard with saturated benzoic acid for concentrations used in procedure.

   NaCl, 0.9%

   Hexokinase reaction mixture
   Prepare according to manufacturer's directions.

6. DNA

   a. Isolation of DNA from *E. coli*
   Buffer for *E. coli*, 0.15 M NaCl, 01 M EDTA, pH 8.0
   Dissolve 8.76 g NaCl and 30.42 g disodium

ethylenediaminetetraacetate in 850 ml deionized H_2O. Adjust pH to 8.0 with 50% NaOH. Dilute to 1 L with deionized H_2O.

SDS, 25%

Add 25 g sodium dodecylsulfate (lauryl sulfate) to 90 ml deionized H_2O. Heat and stir gently to dissolve. Dilute to 100 mls with deionized H_2O.

$NaClO_4$, 5 M

NaOH, 0.5 N

Chloroform:isoamyl alcohol, 24:1 (v/v)

HCl, 0.6 N

Dilute 50 ml concentrated HCl to 1 l with deionized H_2O.

HCl, 0.5 N

Dilute 41.7 ml concentrated HCl to 1 l with deionized H_2O.

Standard saline—citrate (SSC), 0.15 M NaCl, 0.015 M trisodium citrate, pH 7.0

Dilute saline—citrate (SSC), 0.015 M NaCl, 0.0015 M trisodium citrate, pH 7.0

67% ethanol:SSC

Add 66.7 ml 95% ethanol to 33.3 ml standard SSC. Mix.

Nucleic Acid Solution

1). Derived from *E. coli*:

Isolated DNA and RNA from *E. coli* (see p. 50). Rinse the nucleic acids (on glass stirring rods) in a beaker of 67% ethanol:SSC mixture. Place washed nucleic acids in a tube containing 25 ml dilute SSC. Refrigerate 2–3 days. After solubilization, nucleic acid solution can be frozen.

2). Derived from pure DNA and RNA:

Add 1 g pure soluble RNA and 0.6 g DNA threads to 1 liter dilute SSC. Refrigerate 2–3 days. After solubilization, solution can be frozen.

b. Isolation of DNA from whole human blood

All reagents made in double distilled H_2O.

Sucrose-tris-Triton solution

Dissolve 1.211 g tris (hydroxymethyl) aminomethane in about 900 ml of H_2O. Adjust to pH 7.5 with 6 N HCl. Add 109.54 g sucrose and 1.016 g $MgCl_2 - 6H_2O$. Mix until dissolved. Add 10 ml Triton X-100 and mix well. Dilute to 1 liter with H_2O.

Tris-HCl, 0.1 M, pH = 7.5

Dissolve 1.211 g tris (hydroxymethyl) aminomethane in about 90 ml of H_2O. Adjust pH to 7.5 with 6 N HCl. Dilute to 100 ml with H_2O.

EDTA, 0.25 M

NaCl, 2 M

Nuclear lysis buffer

Dilute 1 ml of 0.1 M tris-HCl (pH 7.5), 0.88 ml of 0.25 M EDTA, and 0.05 ml of 2 M NaCl to 10 ml with H_2O.

SDS, 12.5%

Dissolve 12.5 g lauryl sulfate in about 90 ml H_2O. Heat and stir gently to dissolve. Dilute to 100 ml with H_2O.

Proteinase K, 1.75 mg/ml

Ammonium acetate, 8.333 M

Ethanol, 95%

NaOH, 2.5 M

Tris-HCl, 1 M, pH 8.0

Dissolve 12.11 g tris (hydroxymethyl) aminomethane in about 90 ml of H_2O. Adjust pH to 8.0 with concentrated HCl. Dilute to 100 ml with H_2O.

Ammonium acetate, 5 M

NaOH, 0.2 M

7. Glycosylated hemoglobin

Wash buffer

Dissolve 19.27 g ammonium acetate in and dilute to 1 liter with deionized H_2O. Adjust pH to 7.8 with 0.25 M NH_4OH. Dissolve 0.20 g NaN_3 and 10 ml Triton X-100 in each liter of buffer.

Eluting buffer

Dissolve 12.11 g tris (hydroxymethyl) aminomethane in and dilute to 1 liter with deionized H_2O. Adjust pH to 8.5 with 6 N HCl. Dissolve 0.20 g NaN_3, 36.44 g sorbitol, and 10 ml Triton X-100 in each liter of buffer.

Aminophenyl borate Sepharose

Aminophenyl boronic acid is coupled to Sepharose CL-6B by the method of Cuatreacases (Cutreacases, P., and C. B. Afinsen, 1971. Affinity Chromatography. In *Methods in Enzymology*, ed. S. P. Colowick and N. O. Kaplan, Vol. 22, pp. 351-355. New York and London, Academic Press.) Wash 25 ml of Sepharose CL-6B with 300–400 ml distilled H_2O and suspend in about 10 ml of 0.1 M $NaHCO_3$. Add 30–40 ml distilled H_2O to 7.5 g cyanogen bromide (3 g/10 ml Sepharose). Stir cyanogen bromide until dissolved or nearly dissolved (one to two hrs); and then add rapidly to Sepharose. Stir mixture constantly. Maintain pH between 10 and 11 by adding 50% NaOH. Maintain temperature at 20–25°C by adding small amounts of ice to the mixture. After the pH has stopped falling or falls very slowly (about 15 minutes), stop the reaction by filtering the Sepharose on a sintered glass filter funnel. Wash with 1,000 ml cold 0.1 M $NaHCO_3$. Immediately, add m-aminophenyl boronic acid (200 mg in 25 ml of 0.1 M $NaHCO_3$, pH 7.5) to the activated Sepharose. Mix slowly by end-to-end rotation in a 50 ml tube for 16 hours at 4°C. Add ethanolamine-HCl (0.8 g/10 ml Sepharose, dissolved in a few ml of distilled H_2O, pH 7.5) to Sepharose mixture and continue end-to-end rotation for 24 hours at 4°C. Wash with 500 ml distilled H_2O, suspend in 0.02% NaN_3 and store at 4°C.

Note: The Sepharose CL−6B and cyanogen bromide were obtained from Sigma Chemical Co., St. Louis, MO. Ethanolamine-HCl and m-aminophenyl boronic acid, hemisulfate, were obtained from Aldrich Chemical Co., Milwaukee, WI. The chromatograph columns were made from 3.5 ml capacity disposable pipettes (P5214–10, Scientific Products). The top of the bulb was removed, leaving the bottom to be used as the reservoir for the column. A small amount of glass wool was placed in the tip of the pipette to form the support for the resin. The volume of the resin bed was ~1 ml. The columns were supported by plastic cap plugs. (Protective Closures, Inc., 2207 Elmwood Avenue, Buffalo, N.Y. 14214).

8. Steroid hormone formation

NADPH, 1 mg/ml

Dissolve 3 mg NADPH in 3 ml of 0.02 M phosphate buffer, pH 7.4.

Prepare fresh daily.

Steroid standard

Dissolve 50 mg each of progesterone, corticosterone, and deoxycorticosterone in 50 ml methanol. Store at 4°C.

Phosphate buffer, 0.02 M, pH 7.4

Dissolve 17.41 g K_2HPO_4, anhydrous, in 100 ml of deionized H_2O (1 M). Dissolve 13.60 g KH_2PO_4 in 100 ml of deionized H_2O (1 M). Mix 32.2 ml of 1 M K_2HPO_4 and 7.8 ml of 1 M KH_2PO_4 in 500 ml of deionized H_2O. Add 0.493 g $MgSO_4 \bullet 7H_2O$ and dilute to 1,700 ml. Check pH and adjust with stock solutions if necessary. Dilute to 2 liters with deionized H_2O.

SU-4885, 5×10^{-4} M

Dissolve 0.0113 g SU-4885 in 100 ml potassium phosphate buffer, 0.02 M, pH 7.4. Store at 4°C.

Enzyme, 4 mg/ml buffer, acetone powder of bovine adrenal cortex

Place 100 mg acetone powder in homogenizer. Add 25 ml 0.02 M potassium phosphate buffer, pH 7.4, and grind. Keep on ice constantly.

Scintillation fluid

Dissolve 10 g PPO and 1 g POPOP in 2 liters toluene. Mix thoroughly until completely dissolved.

Bovine adrenal cortex, acetone powder

Transport bovine adrenal glands from the slaughterhouse to the laboratory suspended in 0.9% NaCl at 4°C. The following steps should be done at 4°C. Bisect glands from narrow edge and scrape the cortex and medulla material from the capsule. Discard the capsules. Weigh the remaining material and suspend in 1:4 (w/v) of buffer (50 mM potassium phosphate, 0.25 M sucrose, pH 7.4). Homogenize the suspension in 50-ml Potter-Elvehjen glass or teflon tubes. Centrifuge the homogenate at 15,000 rpms for 30 minutes. Decant and discard the supernatent solution. Place 20 volumes of acetone (cooled to −20°C) in a 4-liter flask and begin mixing on a magnetic stirrer. Add the pooled suspension very slowly to the acetone. Mix for 10 minutes and then allow the suspended tissue to settle for 15 minutes. Decant off the supernatent solution. (Optional: Add another 10 volumes acetone, mix, and decant. Add 10 volumes diethyl ether, mix and decant.) Filter the residue and remaining liquid on a Buchner funnel with suction. The acetone powder residue (pale tan color) is air dried and then ground to a fine powder with a mortar and pestle. Store at 4°C in a dessicator.

9. Immunoelectrophoresis

Borate—phosphate buffer, ionic strength 0.05, pH 8.4

Dissolve 4.38 g $Na_2B_4O_7 \bullet 10H_2O$ and 2.14 g $NaH_2PO_4 \bullet H_2O$ in and dilute to 1 liter with deionized H_2O.

Impregnation agar

Prepare a solution containing 0.1% Noble agar and 0.05% glycerine in deionized H_2O. Dissolve by boiling. Store in refrigerator.

Buffered agar

Suspend 1.5 g of Noble agar in 100 ml of borate-phosphate buffer. Dissolve by boiling.

NaCl, 1%

Stain solution

Dissolve 1.5 g Amido Black 10B in 1320 ml of rinse solution.

Rinse solution

Mix together 1000 ml of methanol, 1000 ml of deionized H_2O and 200 ml of glacial acetic acid.

Bromphenol blue, 0.1%

10. Na-K by Coleman flame photometer

Acationox, 1%

Sodium blank

Dilute 1% Acationox 1:50 with deionized H_2O.

Sodium standard, 0.75 meq/l Na^+

Dilute 0.75 ml of 1,000 meq/l NaCl and 20 ml of 1% Acationox to 1 liter with deionized H_20.

Potassium blank

This is same solution as sodium standard

Potassium standard, 0.10 meq/1 K^+

Dilute 1.0 ml of 100 meq/l KCl, 0.75 ml of 1,000 meq/l NaCl, and 20 ml 1% Acationox to 1 liter with deionized H_2O.

NaCl stock standard, 1,000 meq/l NaCl

Dilute 58.454 g of dried, dessicated NaCl to 1 liter with deionized H_2O.

KCl stock standard, 100 meq/l KCl

Dilute 7.45 g of dried, dessicated KCl to 1 liter with deionized H_2O.

11. Cl^- by Chloridometer

Cl^- standard, 100 meq Cl^-/l

Dilute 1,000 meq NaCl/l stock standard with deionized H_2O.

PVA solution

Add 100 ml cold deionized H_2O to 1.8 g polyvinyl alcohol. Heat to boiling with stirring. After the PVA has dissolved, cool to room temperature.

Nitric-acetic-PVC reagent

Mix 800 ml of deionized H_2O, 6.4 ml nitric acid, 100 ml glacial acetic acid, and 100 ml of PVA solution.

12. Alveolar air

Bicarbonate solutions

Prepare the following solutions from dried, dessicated $NaHCO_3$, 20 mM (1.6 g/l), 30 mM (2.52 g/l), and 40 mM (3.36 g/l). Prepare fresh daily.

John F. Van Pilsum earned his Ph.D. in biochemistry at the University of Iowa in 1949. He has taught at the University of Minnesota since 1954 and is now professor of biochemistry in the Medical School.

Robert J. Roon received a Ph.D. in 1969 at the University of Michigan, and is now associate professor of biochemistry in the University of Minnesota Medical School.